A guide to small wind energy conversion systems

British Wind Energy Association

A Guide to Small Wind Energy Conversion Systems

Edited by

JOHN TWIDELL
Energy Studies Unit,
University of Strathclyde

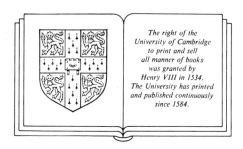

The right of the
University of Cambridge
to print and sell
all manner of books
was granted by
Henry VIII in 1534.
The University has printed
and published continuously
since 1584.

CAMBRIDGE UNIVERSITY PRESS

Cambridge

New York   New Rochelle

Melbourne   Sydney

AAV 2609

Published by the Press Syndicate of the University of Cambridge
The Pitt Building, Trumpington Street, Cambridge CB2 1RP
32 East 57th Street, New York, NY 10022, USA
10 Stamford Road, Oakleigh, Melbourne 3166, Australia

© Cambridge University Press 1987

First published 1987

Printed in Great Britain at the University Press, Cambridge

Library of Congress cataloguing in publication data available

British Library cataloguing in publication data

A guide to small wind energy conversion
systems.
1. Air turbines – Great Britain
I. Twidell, John
621.31'2136   TK1541

ISBN 0 521 26898 2

CONTENTS

EDITORS

General Editor: Dr John Twidell

Subject Editors: Mr Scott Bannister

                       Mr John Evans

                       Mr Peter Fraenkel

                       Mr John Garside

                       Dr David Infield

                       Dr Robert Johnson

                       Mr Bob Leicester

                       Mr Chris Riddell

                       Mr Doug Warne

                       Dr Geoff Watson

Graphics and
illustrations: Mr Peter Hayman

The Editors have worked together as a team to produce this Guide Book.
Individual chapters were drafted by one or two subject Editors, and the
whole text circulated for comments. The General Editor has drawn the
chapters together.   The Editors and Illustrator are all active members
of the B.W.E.A., working in industry, universities and colleges.

THE BRITISH WIND ENERGY ASSOCIATION

(B. W. E. A.)

The aims of the Association are to encourage the development and use of wind power at all scales and for a wide range of purposes. Since its inception in 1978, the Association has become a respected and influential voice to government and commercial enterprise. As part of the European Wind Energy Association, the B.W.E.A. collaborates in developments at international level.

There are now about 500 members of the Association, all with a professional or technical commitment to wind power activities. Details of membership and activities can be obtained from the B.W.E.A. Administrator.

Regular conferences and day meetings are held at national and local level. In addition the Association frequently helps general enquirers including educational establishments. The Association has a range of publications, all aimed at establishing wind power as a recognised and economically viable field of engineering.

Address for enquiries:

The Administrator
B.W.E.A.
4 Hamilton Place
London WIV OBQ

Telephone: 01-499 3515

PREFACE

This Guide Book has been written by members of the British Wind Energy Association to encourage the responsible development, manufacture and use of small wind turbine machines. We do this in the conviction that such machines provide important sources of power and may be cost effective for many users. We have aimed to write in non-technical terms for general readers and to give priority to practical matters. We have always tried to give straightforward and honest answers to the questions commonly asked by potential users of these machines.

Users will include farmers, estate owners and rural householders. Consultant engineers will often be involved in advising about machines and superintending their installation. Our readers should be able to use this book to evaluate the quotations made and the work done. Likewise they will need to know about local planning applications and rating valuations. If electrical grid connection is contemplated, then they must understand the regulations and tariffs regarding the exchange of electrical power. Financial implications will be uppermost in the minds of most users, so we have included details on cost evaluation and payback time.

The UK was at the forefront of wind power development until the 1950s, but funding and official interest declined in the subsequent years of cheap oil. Therefore for modern machines, the UK was a late starter. This is disappointing, because we have some of the most favourable wind regimes of anywhere on earth.

Now however the position has changed, and wind power is considered by governments and industry alike as a 'front-runner' for new power sources. We hope that this book will encourage manufacturers and users to take an active interest in wind power. In particular we hope to see well designed and well engineered machines sold on the open market. This will demand product certification and rigorous safety standards, and the cooperation of public authorities. We hope that this book will be a useful guide to all concerned.

The British Wind Energy Association is a professional body with an international reputation. As such the association sets itself high standards and is careful to remain impartial. This book has been produced by the Association with much care and thought. However, the Association can accept no responsibility for information and data prepared by others and included in the text, nor for any commercial interpretations made from the material in the book or the results of consequent actions. By way of updating information and correcting any possible errors, the Association expects to publish regular articles and news in its magazine 'Windirections', available from the Association's headquarters.

As authors we acknowledge our debt to the previous excellent B.W.E.A. book 'Wind Energy in the 80s', which we strongly recommend to readers of this book. We are indebted to Dr John Burton for information and literature on wind turbine water pumps. Mrs Nicola Lawrence took on the task of assembling our text and typing the draft copy. Her comments and suggestions have been gratefully incorporated by us. The task of typing and arranging the final copy was done by Mrs Evelyn Smith and Mrs Ann Lawson of the Energy Studies Unit, University of Strathclyde. We are indebted to them. Ms Teresa Anderson and Mr Ian Robertson kindly assisted with the production.

We hope that this book will lead to your continued interest in using the power of the wind and that you will become a supporter of the British Wind Energy Association.

John Twidell
Energy Studies Unit
University of Strathclyde
Glasgow G1 1XQ
Scotland

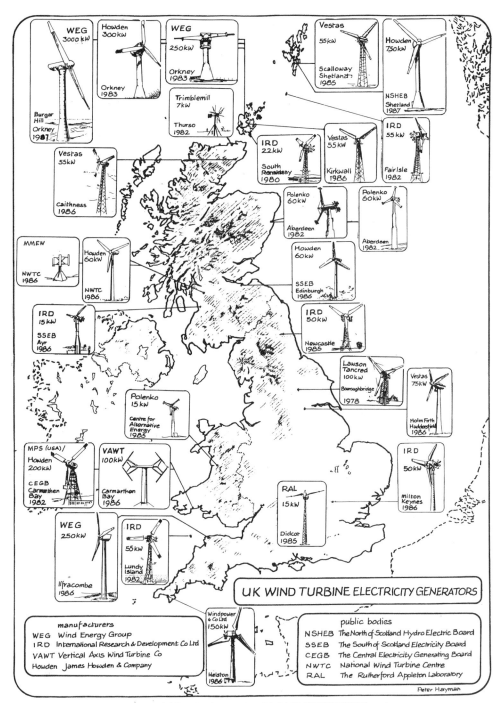

UK WIND TURBINE ELECTRICITY GENERATORS

WEG 3000 kW — Burgar Hill Orkney 1987
Howden 300 kW — Orkney 1983
WEG 250 kW — Orkney 1983
Trimblemill 7 kW — Thurso 1982
Vestas 55 kW — Scalloway Shetland 1985
Howden 750 kW — NSHEB Shetland 1987
IRD 55 kW
IRD 22 kW — South Ronaldsay 1980
Vestas 55 kW — Kirkwall 1986
IRD 55 kW — Fairisle 1982
Vestas 55 kW — Caithness 1986
Polenko 60 kW — Aberdeen 1982
Polenko 60 kW — Aberdeen 1982
MMEW — NWTC 1986
Howden 60 kW — NWTC 1986
Howden 60 kW — SSEB Edinburgh 1986
IRD 15 kW — SSEB Ayr 1986
IRD 50 kW — Newcastle 1985
Lawson Tancred 100 kW — Boroughbridge 1978
Vestas 75 kW — Holm Firth Huddersfield 1986
Polenko 15 kW — Centre for Alternative Energy 1985
MP5 (USA) / Howden 200 kW — CEGB Carmarthen Bay 1982
VAWT 100 kW — Carmarthen Bay 1986
IRD 50 kW — Milton Keynes 1986
RAL 15 kW — Didcot 1985
WEG 250 kW — Ilfracombe 1986
IRD 55 kW — Lundy Island 1982
Windpower & Co Ltd 150 kW — Helston 1986

manufacturers

WEG   Wind Energy Group
IRD   International Research & Development Co Ltd
VAWT   Vertical Axis Wind Turbine Co
Howden James Howden & Company

public bodies

NSHEB The North of Scotland Hydro Electric Board
SSEB   The South of Scotland Electricity Board
CEGB   The Central Electricity Generating Board
NWTC   National Wind Turbine Centre
RAL   The Rutherford Appleton Laboratory

Peter Hayman

PREPARED BY THE ENERGY STUDIES UNIT, UNIVERSITY OF STRATHCLYDE   January 1987

Map of U.K. Wind Turbine Power Generators

1   INTRODUCTION

Contents

1.1   WIND ENERGY

There has been an interest in making wind driven machines for many hundreds of years, mainly using historical designs for grinding grain and pumping water.   However this book is about the modern use of wind turbines and the new designs made possible with the latest engineering techniques.   This modern interest has its roots in the wind machines built in the 1930s, 1940s and early 1950s for generating electricity.   Several countries, including the U.K., had commercial experience of such machines, but the advent of world-wide supplies of cheap oil decreased this interest in wind power, and so many projects were abandoned.   In the 1970s oil prices increased rapidly, and interest has returned to wind power again.   This has been especially noticeable in Denmark and the U.S.A. where there are now tens of thousands of modern machines and many established wind turbine manufacturers.   Other countries also have national programmes for wind energy development, but governments have tended to plan for very large machines to generate several megawatts of power.   At the commercial level however, companies generally market smaller machines for power production capacity between about 10 kW and 300 kW.   Yet other companies have interests in very small machines for charging batteries for lighting at remote installations (see Chapter 4).   With machines of all sizes, the advent of reliable and cheap electronic control enables the machines to be operated automatically and at optimum efficiency. Combined with the use of modern reinforced and laminated materials for turbine blades, these developments enable machines to be economically competitive against traditional fossil fuelled generating sets.

The rapid growth of the wind power industry in Denmark and the U.S.A. has been especially noticeable.   This has been possible because of their central and state governments' strong support for wind energy arising from concern about future oil supplies.   National standards have been established and financial incentives funded.   In this way a market has been created to which industry could respond.   In Denmark most of the machines are at farms and rural communities.   Until recently the machines had an average capacity of about 60 kW, but now machine capacities of about 100 kW are becoming common.   The power is mostly used to sell to the national electricity grid or to substitute for oil fired heating.   About twenty firms manufacture a variety of machines with perhaps 70% of sales going for export. In the U.S.A. the tens of thousands of machines are justified for the electricity they put into the regional electricity grids, thereby earning money for the owners and offsetting their electricity bills.   Growth has been most rapid in

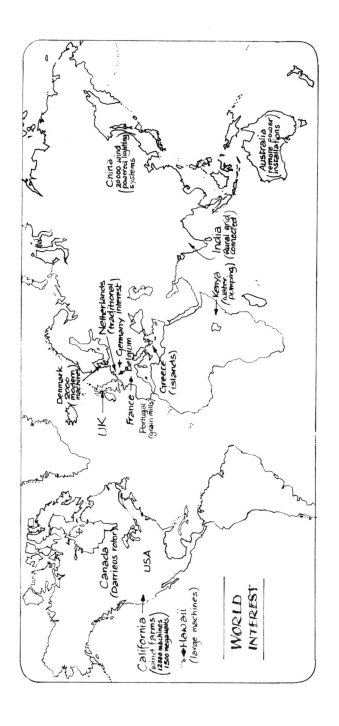

Fig.1.1 World interest in wind turbine generation

California, where there is now much interest in grouping tens to hundreds of machines in arrays called wind farms. These and other countries with considerable experience are all shown on the map of Figure 1.1.

Within Europe there is considerable encouragement from the European Economic Community for wind power generation from small and medium sized machines. Demonstration equipment has been funded on sites throughout Europe, and there is co-ordination of information from test and development centres. It is likely that much future development will be coordinated by E.E.C. offices as the nations of Europe try to establish common regulations and specifications.

1.2   THE WIND RESOURCE
Figure 1.2 is a map of the world-wide wind resource. Winds tend to be strongest in the middle latitudes, especially in maritime regions. A general rule for having a regular and useful wind resource is to be near (or on) the sea, or to be in open country. Winds tend to be strong on hilltops or on mountains, but in these surroundings there is often gustiness and turbulence that can increase difficulties. Good sites exist however on smooth, rounded hilltops where wind speeds may be increased by 30% because of the beneficial contour. As a general rule, wind power is likely to be useful if the average wind speed is greater than 5 m/s (11 m.p.h.), and will be especially attractive for average speeds above 8 m/s (18 m.p.h.).

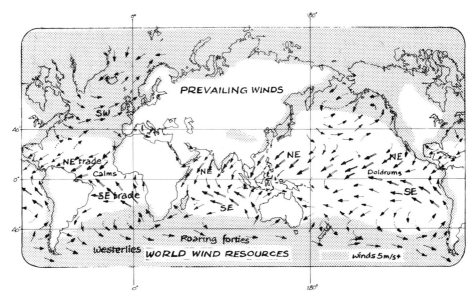

Fig.1.2 Prevailing strong winds. The shaded areas indicate regions of wind attractive for wind power development, with average wind speed more than $5 \times ms^{-1}$, and average generation more than 33% of rated power. Note the importance of marine situations, and beware of non site related generalisations. (Reproduced from Twidell and Weir "Renewable Energy Resources", E & F N Spon Ltd.)

The U.K. has many areas with average wind speeds suitable for electricity generation. Details are given in Chapter 2. The economics of wind power are most favourable where conventional fuel prices, such as oil, are expensive and where there are strong, regular, winds. Such conditions are often found on islands, and so, for example, many wind projects are considered for the Western and Northern islands of Scotland.

The measurement and analysis of the wind resource at a particular site is not an easy task. Fortunately much more information is becoming available for the general public, and instrumentation is becoming cheaper and more common.

Professional assistance is available from consultants and manufacturers to assess the power generating potential of any particular site. Chapter 2 gives much more detail on these aspects.

1.3 BASICS

Wind is an unseen form of energy which may be harnessed for practical application. To do this rotating turbine blades have to be placed in the wind, and the power extracted by generating electricity, turning a shaft or pumping a fluid. Small wind turbines produce power up to about 100 kW, and this power is therefore useful for farms, houses and small-scale industry. Very small machines, producing about 50 - 500 W are useful for battery charging. In this introduction we shall show how estimates of wind power potential can be made and explain the essential features of small wind energy conversion systems. All these points, and more, will be considered in detail in the later chapters.

Typical levels of power derived from the wind are shown in Figure 1.3 and Table 1.1 for three sizes of modern machines. From such analyses, certain conclusions can be made:

** The power in the wind increases with height. At any one height, the power will also depend on the geographical location and the immediate surroundings. For small machines, siting becomes very important.

** Useful power can be produced in relatively moderate winds. For example a fresh breeze of 12 m/s (27 m.p.h.) passing through a turbine of diameter 6 m, the power in the wind is 23 kW. This power rapidly increases in stronger winds, as shown in Table 1.2, where wind speed is related to the well known Beaufort scale.

** A wind turbine has to allow some wind to escape behind the machine, and in practice a modern turbine will extract about 40% of the power in the wind. This efficiency, called the coefficient of performance, is an important factor for the machine, as explained in greater detail in Chapter 3. In our example above, the machine

4

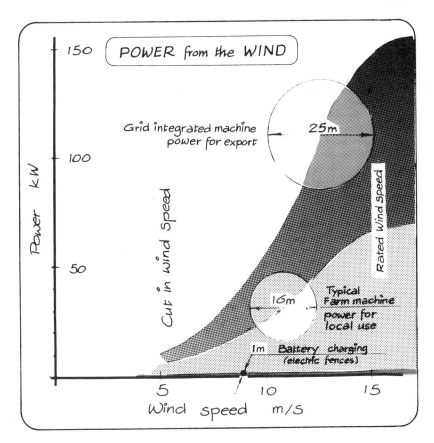

Fig.1.3   Power from wind turbines against wind speed

Rated power of typical machines in wind speeds about 20 m/s.

class	v.small		small			medium			large		very large		
rated power :kilowatts	0.5	1	10	15	50	100	150	250	500	1000	2000	3000	4000
rotor diameter :meters	1,4	2	6,4	10	14	20	25	32	49	64	90	110	130
time per revolution :seconds	0.1	0.2	0.3	0,4	0,6	0,9	1,1	1,4	2.1	3.1	3.9	4,8	5,7

Table 1.1   Rated power of typical wind turbines in wind speeds of about 20 metres per second (44 mph)

Wind scales

Beaufort scale number	meters /second (m s⁻¹)	kilometers /hour (km h⁻¹)	miles per hour (m.p.h.)	knots	Description World Meteorological Organisation (1964)
Force 0	0		0		Calm
	0.4	1.6	1	0.9	
Force 1	1.8	6	4	3.5	Light air
	3.6	13	8	7	
Force 2					Light breeze
	5.8	21	13	11	
Force 3					Gentle breeze
	8.5	31	19	17	
Force 4					Moderate breeze
	11	40	25	22	
Force 5					Fresh breeze
	14	51	32	28	
Force 6					Strong breeze
	17	62	39	34	
Force 7					Near gale
Force 8					Gale
	21	76	47	41	
Force 9					Strong gale
	25	88	55	48	
Force 10					Storm
	29	103	64	56	
Force 11					Violent Storm
	34	121	75	65	
Force 12					Hurricane

Working Winds

Conversion Factors

	m/s	km/h	m.p.h.	knots	
	0.278	1	0.658	0.540	
	0.447	1.609	1	0.869	
	0.515	1.853	1.151	1	
	1	3.6	2.237	1.943	

Table 1.2(a)   The Beaufort wind speed scale

Wind effects

Land	Beaufort number Description	Wind turbines	Sea
	0 Calm 1 Light air 2 Light breeze	non operational	
	3 Gentle breeze		Glassy sea occasional crests
	4 Moderate		White crests common
	5 Fresh	1/3 rated capacity	Crests everywhere
	6 Strong	nominal rated capacity	Larger foaming crests
	7 Near Gale	maximum output	Foam in streaks
	8 Gale	commence shut down or self stalling	Dense streaks
	9 Strong gale 10 Storm	design criteria against damage	Long breaking crests
	11 Violent storm 12 Hurricane	machines automatically shut down or pre collapsed	filled with spray ships hidden

Table 1.2(b) Effects of the wind and the Beaufort scale

would extract 9 kW of electricity from the 23 kW of power in the wind itself. Over a whole year in a windy location, this machine might produce 35,000 kWh units of electricity. Later chapters will go into much more detail on the output and economic assessment of such machines.

** The height of small power generating machines is about the same as domestic buildings, and in general the scale of machinery is similar to agricultural equipment such as tractors and lifting equipment. Table 1.1 gives the approximate size of wind turbine rotors as related to their maximum or rated power output.

** Power production increases extremely rapidly with increase in wind speed. In fact the relationship is such that the power in the wind varies as the cube of the wind speed. The cubic dependence means that a doubling of wind speed produces eight times the power in the wind. It is therefore extremely important to have a windy site and an open location for the machine. In practice mean annual wind speeds of more than 5 m/s (11 m.p.h) are necessary. Such conditions are found over most of the British Isles and the mean average wind speed is usually higher near the sea or in local hilly areas (see Figure 1.4). Some parts of Scotland, Wales and Ireland have mean wind speeds of 8 m/s or more, and are extremely favourable for wind power.

** Power cannot be produced if there is little or no wind. Unfortunately such conditions are common, and so on average a wind machine will only function for about a third of the time and produce an average power of about a third of the rated power. For example a 24 kW rated machine on a reasonably good site will produce an average power of about 8 kW, and may be out of action for up to several days at a time. It is essential to allow for this by some form of alternative power such as exchange with mains electricity supplies, or by energy storage.

** Any particular wind turbine will have a rated capacity which is the maximum power that it can produce over an extended period. Turbine manufacturers usually should design this rated capacity to equal the power produced at about twice the average wind speed.

** The energy in the wind is free, but the machine is not. The owner will be investing money at purchase for a return over the lifetime of the machine. It is therefore important to buy a reliable, well designed and strong machine. Such machines exist, but normal business sense must be shown in deciding what to purchase. A major objective of this book is to guide prospective purchasers into sensible financial decisions.

The conclusions outlined above are fundamental to wind turbine power application and so form the basis for any developments. Readers are urged to consider the points most carefully.

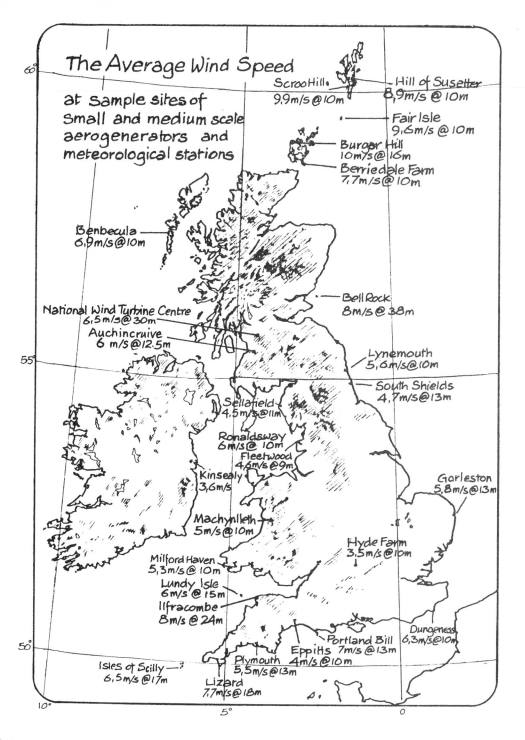

The Average Wind Speed

at Sample sites of
Small and medium scale
aerogenerators and
meteorological stations

ScrooHill 9.9m/s @ 10m

Hill of Susetter 8.9m/s @ 10m

Fair Isle 9.6m/s @ 10m

Buroor Hill 10m/s @ 16m

Berriedale Farm 7.7m/s @ 10m

Benbecula 6.9m/s @ 10m

Bell Rock 8m/s @ 38m

National Wind Turbine Centre 6.5m/s @ 30m

Auchincruive 6 m/s @ 12.5m

Lynemouth 5.6m/s @ 10m

South Shields 4.7m/s @ 13m

Sellafield 4.5m/s @ 11m

Ronaldsway 6m/s @ 10m

Fleetwood 4.6m/s @ 9m

Kinsealy 3.6m/s

Garleston 5.8m/s @ 13m

Machynlleth 5m/s @ 10m

Hyde Farm 3.5m/s @ 10m

Milford Haven 5.3m/s @ 10m

Lundy Isle 6m/s @ 15m

Ilfracombe 8m/s @ 24m

Dungeness 6.3m/s @ 10m

Portland Bill 7m/s @ 13m

Eppitts 4m/s @ 10m

Isles of Scilly 6.5m/s @ 17m

Plymouth 5.5m/s @ 13m

Lizard 7.7m/s @ 18m

Fig.1.4 Average wind speed at sample sites of small and medium scale aerogenerators, and at meteorological stations.

9

1.4 TYPES OF MACHINES

There are many types of wind machine and some examples are shown in Figure 1.5. The principal classification is between turbines with horizontal rotating axes and those with vertical rotating axes. At present the great majority of commercial machines are horizontal axis types. The type of use determines the number of blades on the rotor.

** Turbines with many wide blades start easily, even in a light wind, and produce a large turning force suitable for pumping water or rotating a shaft slowly. Such many-bladed machines cannot turn rapidly however, and are therefore not suitable for generating electricity.

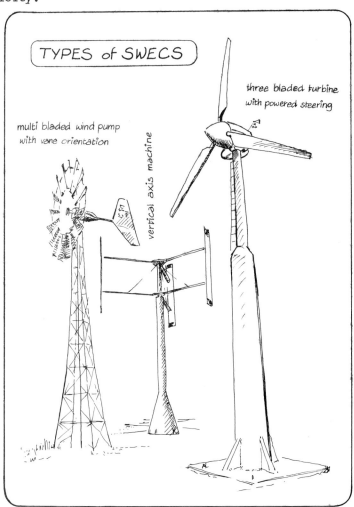

TYPES of SWECS

three bladed turbine with powered steering

multi bladed wind pump with vane orientation

vertical axis machine

Fig.1.5   Types of SWECS

**   Turbines with a few blades, usually two or three, can reach high rotational speeds and are used for generating electricity. Efficient machines for this purpose will have blades designed to high standards of engineering. A very small machine needs to rotate rapidly to generate power, whereas a larger radius machine need not rotate so rapidly at the same wind speed. Many people find that the slower rotational speed of larger machines gives the machines aesthetic appeal and makes them quieter.

**   Horizontal axis machines may have the turbine blades rotating upwind or downwind of the tower. Upwind blades require the machine to have a powered steering system, a tail, or sideways 'fan tail' rotors to maintain orientation into the wind. Downwind blades are normally maintained in position with respect to the wind without such steering systems. A disadvantage of the downwind configuration however is that the tower deflects the wind so causing a wind shadow and unsteady power production.

In addition to the obvious differences in rotor configuration, there are many other differences in internal design between machines. These will be explained when the need arises in later chapters.

1.5   FINANCIAL AND INSTITUTIONAL SUPPORT

   It is not easy for an individual to plan, purchase and operate a wind turbine for significant power generation. Fortunately there is an increasing body of support available to advise on obtaining grants and to give technical advice. Nevertheless expert and knowledgeable assistance will be required to establish a successful installation. Chapters 5 and 6 have been written to give down-to-earth advice on such matters. This will be particularly needed if an individual contemplates buying and selling electricity to the grid, or hopes to sell power to other consumers. Other matters to consider are local rates and taxes, planning permission, grant applications and insurance. The list may seem long, but in practice the task is not too difficult. This Guide Book has been written precisely to answer such questions. Remember that when you buy a machine, normal business sense is required. Check that the seller is authentic and is following good engineering practice. The onus is on the seller to demonstrate to you that his product is acceptable and of proven value. It is always sensible to ask where you can see one of the machines in operation and to talk to established customers. Be particularly careful if the machine is a new product.

The U.K. Energy Act (1983) gave the legal framework for important developments for electricity production and distribution from wind power and other private generation. A private owner has the right to buy and sell power by interchange from the mains electricity grid, and also to transmit his own power by means of the grid. The precise financial and technical arrangements are controlled by regulations stemming from the Act, and details are discussed in later chapters and in Appendix A3.

1.6   TEST CENTRES AND CERTIFICATION

It is extremely difficult for a private consumer to be able to judge the standard and worth of a commercially available wind turbine. For this reason, test centres have been established in several countries to provide advice to both manufacturers and consumers, and to certify equipment and systems.

The U.K. National Wind Turbine Centre is operated by the National Engineering Laboratory at East Kilbride, near Glasgow. This centre co-operates in a European Community network of such centres. It is important that consumers obtain the information produced from these official centres regarding any equipment being purchased.

2   WIND CHARACTERISTICS

Contents

2.1   INTRODUCTION

The energy in the wind is proportional to the cube of wind speed. Since the economics of a wind turbine installation will depend upon the annual energy production that can be achieved, it is very important to find a site which offers the highest overall wind speeds. It is also important to avoid locations with excessive gustiness or turbulence, since this creates high fluctuating stresses on the machine leading to rapid failures due to fatigue. The siting should also take account of the prevailing wind direction to have exposure in that direction. The variation of wind speed on a diurnal (daily) and seasonal basis may also be important. For example, in designing a remote power supply the size of energy storage will depend critically upon matching wind power and load patterns.

For preliminary estimates of wind power production it is therefore important to understand the nature of the wind. It is then possible to make approximate predictions of the conditions at a prospective site, without recourse to expensive long-term measurements with anemometers. For small wind power installations the cost of sophisticated wind measurements could easily exceed the cost of the installation itself, and so cheap methods of estimating wind behaviour are necessary.

The following sections are a brief introduction to the nature of wind and give guidance on suitable site selection.

2.2   IMPORTANT CHARACTERISTICS OF THE WIND

Wind is caused by atmospheric pressure differences which arise from unequal heating of the earth's surface by the sun. At high altitudes, where surface friction has little effect, and when the weather is not changing rapidly, the wind speed and direction are determined by these pressure gradients and by the forces produced by the earth's rotation. These large-scale or synoptic factors result in the major wind zones that we know as the doldrums (around the equator), trade winds (in lower latitude ocean areas), and maritime or temperate conditions in higher latitudes (see Figure 1.2). Nearer the earth's surface (in the lower 1000 meters or so) other forces caused by surface friction, local temperatures and topography come into operation.

For wind power generation from small and medium machines the most important characteristics are those of 'surface' winds up to about 30 metres from ground level.

We need to know:

* Annual mean wind speed
* Daily variations
* Directional distribution
* Variation with height above ground
* Turbulence intensity

Each of these features has a direct bearing on the behaviour and performance of a wind turbine, and most have an influence on where (or where not) to site a machine.

The most important characteristic is annual mean wind speed, since this is the single feature which most influences the annual energy production of the machine at a given site. Long term measurements have been made at numerous stations mostly operated by the Meteorological Office. Instruments may be used to produce a continuous record of wind speed and direction on moving paper charts. Maps of annual mean wind speed may be produced from the results but there are difficulties in the presentation and interpretation of national wind maps since the wind in any particular locality is strongly influenced by surrounding features, such as hills, which are impractical to represent on a national scale. Therefore large scale wind maps may be misleading for a particular site. Several attempts have been made to produce maps of wind speed and Figure 1.4 is an example using data from wind turbine installations.

The annual mean wind speed is an average over the whole range of different wind speeds throughout the year, and the compositon of this range of wind speeds is also important. This is generally described by a wind speed duration curve, which gives the proportion of time for which the wind speed reaches or exceeds certain levels. An example of a wind speed duration curve is shown in Figure 2.1a. This curve can be combined with the power performance curve of any proposed machine (Figure 2.1b) to convert wind speed to power output. The resulting curve (Figure 2.1c) is generally known as a power duration curve. The area beneath this curve corresponds to the predicted annual energy output of the machine. It can be appreciated that the shape of the wind speed duration curve, as well as the shape of the power performance curve, has a strong effect on the annual energy production.

The variations of wind speed with time are predominantly random, although of course they are influenced by the passage of weather systems on a general (synoptic) scale. It is difficult to describe a pattern which has any statistical significance or validity. An average of the hour by hour pattern can be derived from long-term records, and an example is shown in Figure 2.2. Even the general shape of this pattern will not be consistent from area to area, depending for instance upon the influence of land and sea breeze effects. The patterns will also be subject to very wide variation on a day to day basis, and it cannot be relied upon in any simple fashion for the sizing of energy storage to accompany a wind energy system, or generally for the matching of a wind turbine to a time-varying load.

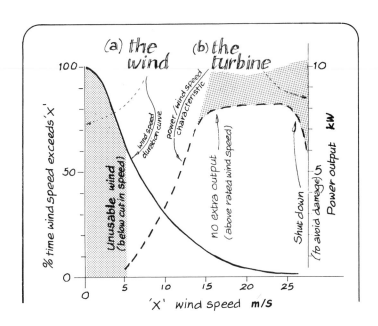

Fig.2.1a   Wind speed exceedance curve

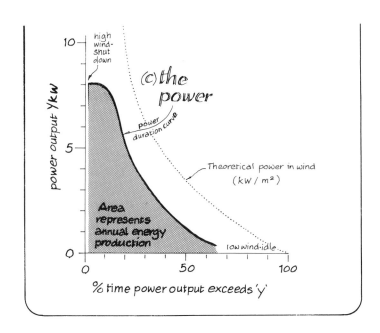

Fig. 2.1b   Wind turbine power as a proportion of time
:   power duration curve

Directional characteristics of the wind at any site are influenced mainly by large-scale weather patterns prevailing over the U.K. The dominant direction is southwesterly (from the south west), except in a limited area along the east coast, where easterly winds off the North Sea prevail. The relative importance of winds in particular directions is customarily illustrated by a wind-rose. This shows the percentage occurrence of winds in the eight major compass points, and may also include the broad composition of wind speeds in each of these eight directions. An example of a wind-rose is given in Figure 2.3; this gives a good idea not only of the prevailing direction, but where the strongest winds, and therefore the most energy, arises. Although the large-scale southwesterly conditions influence most of the country, local effects can affect direction in the same way that they affect wind speed. This can be especially pronounced if a prospective wind turbine site is in a valley or adjacent to a large feature such as a hill or lake.

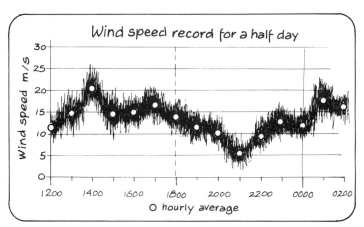

Fig.2.2   Wind speed variation during a day

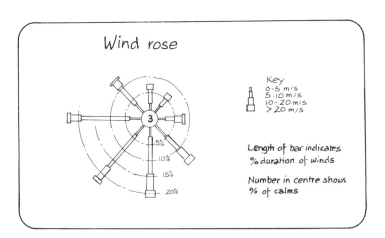

Fig.2.3   Wind rose

16

The variation of wind speed with height above ground (often called the wind shear, 'vertical gradient' or 'vertical profile') and turbulence are important and related features. Brief reference was made earlier to high altitude winds, and the effect of surface friction which slows the wind down nearer the ground. These mechanisms result in vertical profiles of wind of the general type shown in Figure 2.4. The vertical profile is described approximately for wind energy applications using an exponential form;

$$v/u \quad = \quad (h/z)^a \qquad \text{(Equation 2.1)}$$

where v is the wind speed at height h above the ground, in relation to the wind speed u at a standard height z. Most standard meteorological measurements are taken at a height of z = 10 metres. Because the power law exponent 'a' represents a simplification, it is not strictly constant for any particular site, and varies with direction, wind speed, the height range under consideration and atmospheric conditions generally. However, average figures can be attached to certain types of terrain, and Table 2.1 indicates the values to be expected. Equation 2.1 is used when the hub height of the wind turbine is different from the height at which any wind measurements have been made, or when a taller tower is being considered to improve the output of the machine.

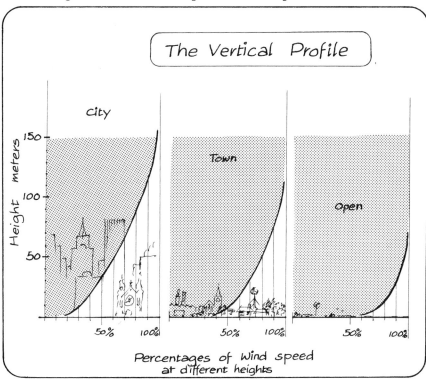

Fig.2.4   Percentages of wind speed at different heights

Wind speed terrain index with height

	index "a"	
Mud flats Ice Snow Smooth sea Sand	0,10	
Low / mown grass Steppe	0,13	
Fallow field high grass Low scrub	0,19	
Forest - Woodland Suburb - City	0.32	

Table 2.1   Wind speed index with height

Surface roughness characteristics and obstructions are also responsible for the generation of turbulence or 'gustiness'. In strong winds, mechanical mixing generated at the surface is the dominant turbulence producing mechanism. A rough surface therefore, such as woodland and built-up areas, will generate higher turbulence than a smooth surface such as grasslands, meadow etc.

The changes in wind speed and wind direction are associated with the natural turbulence of the air motion. These effects are easily recognised in water moving rapidly over a rough stream bed, but are difficult to comprehend in air unless smoke is injected into the fluid motion. Nevertheless the effects are extremely important for wind turbines, and perhaps especially for small machines so often sited near local obstructions. Increased turbulence in the wind is detrimental to the performance of wind turbines, particularly regarding the quality of power output and stresses on the turbine. Generally, therefore, the greater the level of turbulence, the less suitable the location. The overwhelming complexity of turbulence renders a straightforward and comprehensive description impossible. The general measure of turbulence is turbulence intensity, equal to the standard deviation of wind speed divided by average wind speed for the same set of measurements.

During its lifetime, a wind turbine will experience wide extremes of wind speed from storm gusts. These gusts require special consideration since they represent the maximum force which the machine will be required to withstand. It is clearly important to determine a realistic value for the extreme wind since an overestimate could result in a costly overdesign and an underestimate could result in a failure. The underlying difficulty in estimating an extreme wind is that no matter what value is used, there is always a chance that it will be exceeded. The estimation of extreme wind characteristics should be regarded as a fundamental starting point in the treatment of wind loads on structures. Consumers should be able to receive advice on such matters from the National Wind Turbine Centre or local meteorological office.

2.3   SITING A WIND GENERATOR

It is generally agreed that the ideal position for a wind generator is a smooth hilltop with a long, flat uninterrupted exposure, at least in the prevailing direction see Figure 2.5a. Near the top of such a hill the wind speeds will be increased by perhaps 50% over speeds at the same height above ground before the hill. This is called the 'speed-up' effect. A good position will offer a high annual mean wind speed, giving maximum energy production with a favourable vertical profile and low turbulence levels, thus minimising the transient and cyclic loadings on a machine and the risk of fatigue damage.

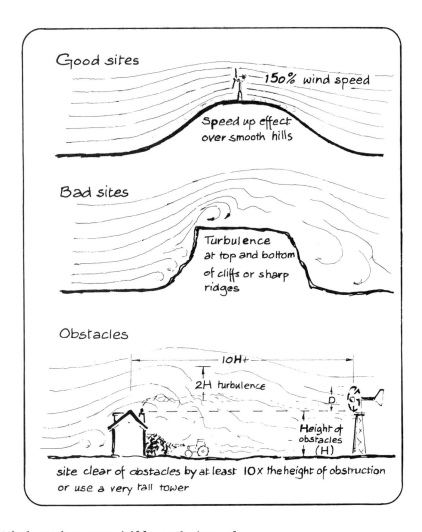

Fig.2.5   Wind motion over hills and obstacles

In practice few perfect sites exist, and the positioning of small wind turbine generators seldom permits adequate flexibility to make optimum use of beneficial conditions. Local barriers, obstructions or special topographical features of some sort will usually be present, and it is difficult to provide guidance on their detailed effects. Several documents and reports have been provided for assistance on this subject, and one of the more useful is the Siting Handbook for Small Wind Energy Conversion Systems produced by Battelle Pacific North West Laboratory (see Bibliography, Wegley et al., 1980). It is however most important to avoid highly turbulent sites near sharp obstructions, Fig.2.5.b.

A wind turbine generator must be positioned well clear of any obstructions to a clear wind. A good rule of thumb is to place the turbine at a distance from such obstacles of at least ten times the height of the obstacles, or to use a tower that lifts the turbine to at least twice their height, Fig.2.5c.

2.4. PREDICTING AND MEASURING WIND CONDITIONS

The earliest and simplest form of wind measurement was based upon observation, giving rise to the Beaufort scale of wind speed. The specification for the Beaufort number 1 to 12 for inland areas is given in Table 1.2 of section 1.3. The simplest form of wind speed prediction for a site is to note the conditions over a period of several months, if possible at hourly intervals. If contact can be established with the local meteorological station, the recorded hourly wind speed at that station can be obtained, and by comparing these with the observations at the intended site some general idea of the relationship of the winds at the two places can be obtained. The long term mean wind speed for the meteorological station will be known, and that for the prospective wind turbine site can therefore be estimated, very approximately.

An indication of local wind strength and direction can often be obtained by observing the deformation of vegetation and trees, Fig.2.6. Such so-called natural indicators can provide reliable information on the long-term average wind properties. A description of vegetation as an indicator of high wind velocity is given by Hewson et al (see Bibliography) and is also referred to in Wegley et al (1980). The absence of isolated trees in an ungrazed area is often a clear indication of strong winds. It must be recognised however that such techniques are only qualitative.

More definitive and reliable data can of course be obtained by purchasing or hiring an anemometer. Until recently, the cheapest and most reliable form of anemometer was a 'run-of-wind' instrument which embodies a counter (odometer) to register the number of revolutions of the cups. The reading given is in the form of kilometres or metres of wind which have run past the anemometer. Reading the counter at frequent intervals, (say for 10 minutes every three hours, daily or weekly) will therefore give the average wind speed over that period if the run-of-wind is divided by the period. Such readings can then be compared with records from the nearest meteorological station, or if taken over a long enough period (at least a year is usually necessary) they can be used in their own right. Correction to the intended tower height for the wind generator according to Equation 2.1 for vertical gradient will usually be necessary. A run-of-wind anemometer and mast can be obtained at a cost of about £250 to £400.

Traditional recording anemometers are more useful if the time varying pattern of the wind speed is important, but facilities for analysis of the records are usually necessary. Specialised instruments such as fast response anemometers for turbulence measurement or kites for investigation wind speed vertical gradient are also useful for surveying purposes.

Wind measurement techniques have been significantly improved in recent years by the development of microprocessor based systems. The development has now reached a stage where an anemometry system can be set up and left, unattended, for several months to a year of data collection. Soon semi-manual methods of recording wind data will be outdated. The microprocessor read-out should give the wind speed distribution, such as in Fig 2.1.a.

Fig. 2.6 Vegetation affected by the prevailing wind

A general schematic of how a modern wind measurement system is configured is shown in Figure 2.7. This system shows an anemometer in use. In principle other devices such as a wind direction vane, temperature and atmospheric pressure transducers can be readily included, however these are not essential for general site evaluation studies.

The fundamental application of the microprocessor is for data recording and immediate analysis. Greatest impact has been made on the measurement and recording of mean wind speed and wind duration, and power prediction. Techniques for turbulence measurement, although dramatically improved by the microprocessor, remain complex and still require specialist skills, both in terms of data collection and data interpretation. Since power prediction depends on the cube of the wind speed v^3, immediate analysis of average values of v^3 over specified intervals is important.

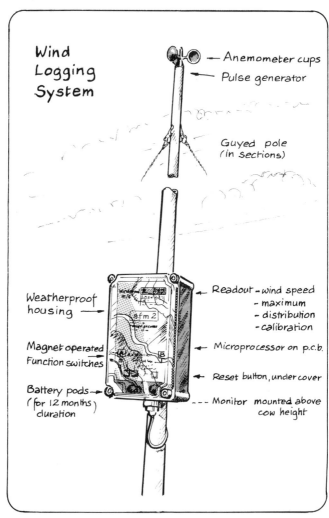

Wind Logging System

— Anemometer cups
— Pulse generator

Guyed pole
(in sections)

Weatherproof housing —

Readout - wind speed
   - maximum
   - distribution
   - calibration

Magnet operated Function switches

— Microprocessor on p.c.b.

— Reset button, under cover

Battery pods —
(for 12 months)
duration

--- Monitor mounted above
cow height

Fig 2.7 A microprocessor based anemometry system for wind logging

For the measurement and recording of mean wind speed characteristics there are an increasing number of anemometry systems available on the market. The following features are important:

** Low power consumption. Clearly, it is desirable that battery changes should be kept to a minimum. This is especially important if the measurement site is not easily accessible. Minimum battery life of several months can be expected and some systems operate unattended for a year. With the rapid developments in microprocessor application this aspect is improving all the time.

** On-line analysis. Until recently wind data logging systems recorded only the number of rotations of an anemometer over many intervals of about ten minutes to an hour. Such systems accumulate vast amounts of data that are difficult to analyse. The latest systems analyse data immediately in terms of 'running averages' and distributions of wind speed. Such on-line analysis dramatically reduces the effort for analysis and produces results immediately.

** Variable averaging time. Logging systems that do not have on-line analysis record the number of anemometer rotations over set 'averaging times'. The recorded numbers remain on some form of data logger. Most loggers have averaging times which are factory fixed, e.g. 10 minute averages only, or 60 minute averages only etc. However the ability to vary the averaging time is very beneficial. A useful range of averaging times would be from, say, a minute to an hour. It must be noted though, that the shorter the averaging time, the sooner the logger memory fills up.

** Transfer capability. If the system does not have on-line analysis, it is essential to be able to transfer the logged data from the data logger to some other device such as a portable microcomputer. This allows the logger to remain in service continuously and enables off-site data analysis. It is convenient if the transfer visits can be timed to coincide with battery changes.

These features represent only fundamental logging operation and procedure. Other features such as the ability to withstand extremes of temperature and humidity are also vitally important for long-term deployment. General guidance on this cannot be given because the requirements to withstand severe environments are determined exclusively by where the system is to be used. The only advice which can be given is to make sure that the supplier of the system is fully aware of the end use.

3   CHOOSING A MACHINE

Contents

3.1   INTRODUCTION

Wind is variable in nature and wind turbines produce mechanical power only when the wind is available. This mechanical power may be used in one of three ways:

**   It can be utilised immediately as mechanical energy, e.g. water pumping, stirring a fluid to produce heat;

**   It can be converted to electrical power and used immediately either locally or transmitted through a grid network to a remote point of use, e.g. sale of electricity, electrical heat;

**   The energy can be stored in various forms and used in the future, e.g. electric vehicle power packs, batteries for lights.

To meet continuous demands for energy, back-up systems of storage or standby supplies will be required when the wind speed drops below that needed to meet the demand. The basic application will to some extent determine the type of wind turbine required.

3.3   ASSESSMENT OF NEEDS

Having decided the applications for which wind turbines will be required to provide energy, the next step in choosing a machine is to assess the energy need.   For this it is necessary to establish the following parameters:

** Annual gross energy requirement.

** Duration, timing and size of peak demand, minimum demand and continuous demand.

** Type of energy required, whether simple mechanical, heat or electrical.

** Matching between wind energy availability and demand.

** Possibilities of load management to even out demand.

** Percentage of demand which can economically be met by wind turbines.

** Provisions to make up shortfall, or to distribute or store any excess.

** Back-up systems which might be required, and which are available.

Whilst some indication of machine size for various applications has been given in Table 3.2, it is possible to calculate the output of any wind turbines on any site from manufacturers' literature or from first principles using the following formulae:

Available power (Pa) in the wind at any moment is given by:

$$Pa = 0.5 \, d \, A \, v^3 \qquad\qquad \text{(Equation 3.1)}$$

where d is the air density, A is the rotor area and v is the wind speed. The useful power which can be extracted by the turbine rotor is given by multiplying the available power by a coefficient of performance Cp for the particular turbine.   This Cp has a maximum theoretical value of 0.59.   Values of around 0.4 are achievable in practice.   Thus the power output (Po) of a wind turbine rotor is:

$$Po = Cp \ (0.5 \, d \, A \, v^3) \qquad\qquad \text{(Equation 3.2)}$$

A wind turbine will be given a rated designated power by the manufacturer.   Usually this is the maximum power for continuous operation in strong winds, and various measures will be taken to prevent the power rising above this in even stronger winds.   Thus machines are quoted with a rated output power Pr in a rated wind speed vr.

The relationship between the wind speed at which the turbine achieves

29

its rated power Pr, and the mean wind speed vm at the site is also important. If the rated wind speed is high, the machine will seldom achieve full output, conversely if it is too low much of the energy available in the wind will be wasted. Typically the ratio between vr and vm should lie between 1.5 and 2 to maximise the benefits of the installation.

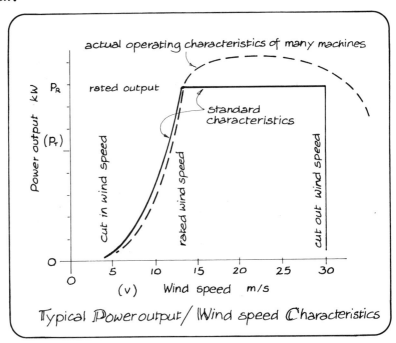

Fig.3.4 Typical wind turbine power output/wind speed characteristic

3.4 MECHANICAL DESIGN OPTIONS
3.4.1 Rotor configuration
There is a great variety of wind turbine designs. They can be classified by a number of features, the main division being the orientation of the axis about which the blades rotate. The conventional looking type is the horizontal axis machine, where propeller-type blades turn machinery on top of a tower. Alternatively, the blades may rotate about a vertical axis (see Figures 1.5 and 3.2). The two main advantages of the vertical axis are that the gearing and generating machinery can be at ground level, and the machine can accept wind from any direction without yawing. Offsetting this is an inherent tendency for uneven rotation as the turbine blades move across the wind each cycle, and for a large number of vibrational modes in the structure.

In horizontal axis types, if the wind passes through the rotor before impinging on the supporting structure, it is termed an upwind design. Alternatively, the rotor can be operated downwind of the tower, in the 'tower shadow'. The advantage of the latter is that no special means are required to hold the rotor into the wind. The disadvantages are higher cyclic loads on the blades and structure, less smooth power output and increased noise as the blades pass through this tower wake.

The number of blades on the turbine has important effects on performance. Turbines with many blades start rotation easily even in a light wind, and so have a high starting torque and low cut-in speed. They are useful for water pumping and frequent operation. However in moderately strong winds the many rotating blades cause uneven flow of the wind through the turbine, and there is a rapid decrease in efficiency.

Turbines with few blades can rotate rapidly in strong winds without causing a disturbance to the air flow pattern, and so these are best for efficient energy capture for electricity generation. They do not start rotating easily in light winds; however at these wind speeds there is less energy to capture in any case.

In practice SWECS for high torque and frequent operation (e.g. water pumping for cattle troughs) have many bladed turbines. SWECS for electricity generation have few blades, with 2 or 3 blades being the most common.

It follows from this discussion that every turbine has an optimum rotational speed with respect to the wind speed. This is expressed by an important factor, tip speed ratio:

$$\text{Tip speed ratio } R = \frac{\text{speed of blade tip}}{\text{speed of wind onto the rotor}} \qquad \text{(Equation 3.3)}$$

The tip speed ratio is a ratio, having no units of dimensional measurement. It incorporates all the important factors of wind turbine rotors, namely radius of rotation, angular speed of rotation and wind speed. For wind turbine engineers it is a crucial non-dimensional performance characteristic for the optimisation of rotor performance. Always ask people selling a wind turbine what the optimum tip speed ratio is. If they do not know, be warned about their credibility!

Figure 3.5 shows typical characteristics for the fraction of power extracted (the power coefficient Cp) as plotted against tip speed ratio. Note that few bladed-rotors operate at high rotational speed for a given wind speed as compared with many-bladed rotors.

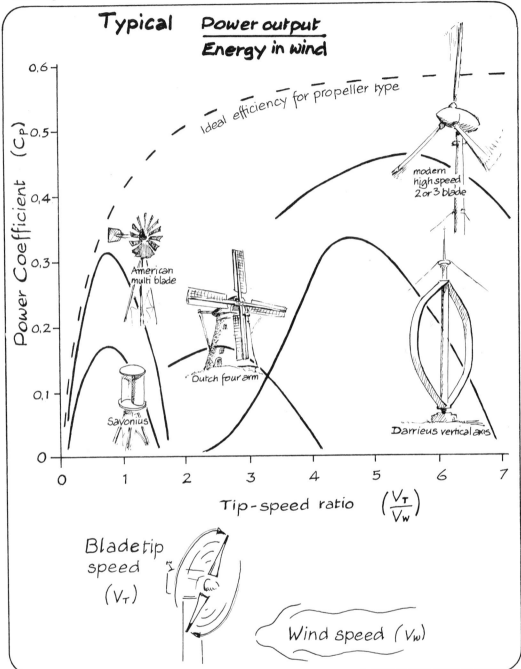

Fig.3.5 Power coefficient versus tip speed ratio for different wind turbine rotors.

Methods of aerodynamic control of the rotor are essential to prevent damage in high wind speeds. There are three general methods:

** **Stall regulation.** The blade design is such that the turbine becomes increasingly inefficient at high wind speeds. Since less and less a proportion of wind is captured, the turbine eventually maintains a near-maximum power output even though the wind speed becomes much greater than the rated speed.

** **Rotor orientation.** A mechanism is incorporated in the tail of upwind turbines so that the whole rotor turns out of the wind at high wind speed. This method is common with very small machines up to about 3 kW capacity and diameter about 4 metres.

** **Blade pitch adjustment.** At high wind speed the whole of the blade or the end tip of each blade is rotated to cause stall conditions. This makes energy capture inefficient, so the rotor stops accelerating and can be made to stop rotating if necessary. Blade pitch adjustment can in principle give excellent rotor control, but may lead to increased cost and failure.

The rotor of a horizontal axis turbine has to turn into the wind as the wind changes direction. Downwind rotors do this automatically, but upwind rotors need a turning mechanism. There are three general methods for this yawing mechanism:

** **Tailfin.** A downwind tail pulls the horizontal axis into the wind. Care has to be taken to prevent an oscillating motion in and out of the wind.

** **Fan tail rotors.** Small side rotors operate on an axis perpendicular to the wind direction. When the main axis becomes misaligned with the wind, these side rotors will rotate to turn the main axis into the wind. Once aligned, the side rotors cease rotation.

** **Powered rotation.** For machines of more than 50 kW capacity it is common to use an electric motor to turn the turbine into the wind. This motor will be activated by a small wind vane in the nacelle of the machine. Obviously a rotor brake has to operate rapidly if electric power fails, since the turning mechanism depends on electric power.

3.4.2 Tower

Using a tall tower to reach into higher wind speeds enables the rotor to capture more energy than a short one, but tower cost is greater. A sensible balance has to be struck and generally it is sufficient to ensure that the tower is tall enough to take the whole of the rotor area out of the terrain induced turbulence, e.g. above tree height, as discussed in Chapter 2, Section 2.3.

A lattice tower uses metal most efficiently, and has lowest cost, but is often less aesthetic than a smooth tubular design. If the machine is large enough, a tubular tower can give complete protection for access and machinery.

3.4.3 Nacelle machinery

The equipment, framework and facilities, placed on a turntable at the top of the tower on a horizontal axis machine, is all termed the nacelle. The main components are the rotor bearings, drive shaft, shaft brake, gearbox, generator, steering mechanism, control components and power output cable, Figures 3.6. In addition facilities must exist for access and for maintenance, using a hinged cover and walkway.

A small rotor can be supported on the gearbox shaft, but usually rotors will have separate bearings. The latter arrangement takes more space but makes maintenance much easier. An epicyclic gearbox is more compact than a spur gearbox, but it is more complex for maintenance. A flexible coupling or fluid coupling in the drive line will smooth any power fluctuations due to wind gusts and torsional oscillations.

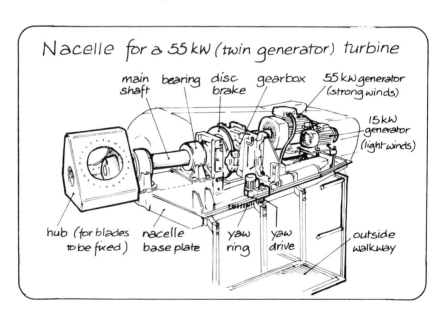

Fig.3.6a   Nacelle for a typical farming or rural site

Braking can be achieved in two basic ways; aerodynamic and mechanical. Included in the first is the use of rotatable blade tips, spoiler devices (such as flaps in the blade surfaces) and complex blade pitching. Turning the rotor out of the wind with a yaw mechanism can also be included in this category.

Purely aerodynamic braking cannot be relied on to bring any rotor to a dead stop and so mechanical brakes are needed. They are also used as parking brakes, and as emergency brakes capable of stopping the rotor from overspeeding. Brakes applied at the rotor shaft prevent braking forces being transmitted through the gearbox, and prevent backlash in the gearbox from the rotor. Note that national design standards for wind turbines will almost certainly require machines to have two independent methods of braking.

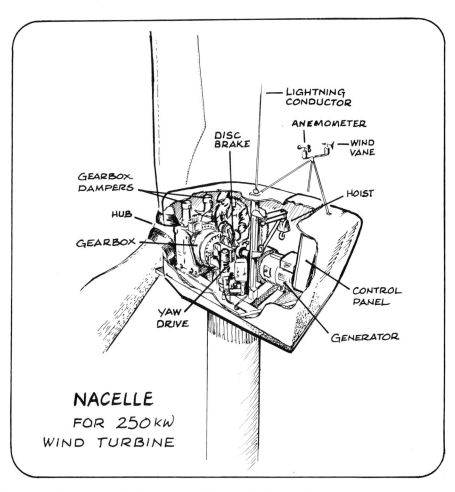

Fig. 3.6b   Example of nacelle layout for a wind turbine for the commercial sale of electricity

3.4.4 Structural materials

Some commonly used materials for various structural components are:

- for the blades:
  Steel, aluminium, wood laminate, glass reinforced plastic
- for the hub:
  Cast iron, fabricated steel
- for the bedplate of the nacelle:
  Steel
- for the nacelle cover:
  Sheet steel, glass reinforced plastic, aluminium
- for the tower:
  Steel, concrete

The blades are subject to a gravity-induced fatigue cycle every rotation. Steel is well documented for its fatigue properties and can be used for the blade spar, usually bonded into a glass reinforced plastic shell. This nevertheless gives a heavy rotor. Blades made entirely from glass reinforced plastic can be used; fatigue properties may be better than steel, but rotors are still heavy. Wood laminate blades are light and appear to have excellent fatigue properties, so this material may be particularly suitable for wind turbine blades.

Whatever material is used for the blade, the blade to hub connection is critical and needs careful design. It is at the root of the blade that the largest forces exist and there must be provision for replacing damaged or worn blades.

3.4.5 Design features

Detailed design is very important since many wind turbines will be in harsh environments. A common fault is for moisture to gather in inaccessible parts causing blades to become unbalanced and corrosion to occur. Most of these difficulties can be overcome, as on motor vehicles, by experience and good design. Many other design features relate to safety and standards of performance. Fortunately international standards are set by test and development centres so design standards and experience are always improving.

Probably the greatest difficulty for wind rotor design is to avoid fatigue induced failures. During a twenty to thirty year lifetime a rotor blade will experience many more stress cycles than an aeroplane wing, so this is a new challenge to technology. Fortunately modern material structures, especially laminates, are able to meet this challenge. Other fatigue related effects in the overall structure and tower can be lessened by careful analysis using modern computer aided design.

3.5 ELECTRICAL DETAIL

A number of different systems of electrical connection for SWECS machines can be used, depending on the type of application. Some of the more common systems are:

** Very small stand alone systems, about 100 W capacity, for battery charging in telecommunications applications (see Chapter 4).

** Stand alone small machines, about 5 kW, generating at varying frequency for non-grid connected applications.

** Stand alone medium sized machines, of about 50 kW capacity, generating usually at fixed frequency for non-grid connected applications, such as remote wind/diesel sets.

** Mains grid connected, medium sized machines, operating at nearly fixed speed.

** Mains grid connected variable speed machines (however these are unusual).

These different systems are described in the following sections and diagrams, apart from the very small systems covered in Chapter 4.

3.5.1 Stand alone power systems

These systems are usually designed to provide a fixed frequency and voltage to loads which are not connected to the power utility grid. The power used as electricity should normally be of acceptable mains quality standard.

The system of Figure 3.7 utilises an a.c. alternator charging batteries via a rectifier. The output from the battery bank may be converted to alternating current at a fixed voltage and frequency by an invertor. A self-excited induction generator could be used to replace the alternator. The batteries in this system must have the capacity to provide the full requirements of the load during extended periods of calm, and so in practice only essential electrical services, e.g. lights, can be supplied because of the high cost of batteries.

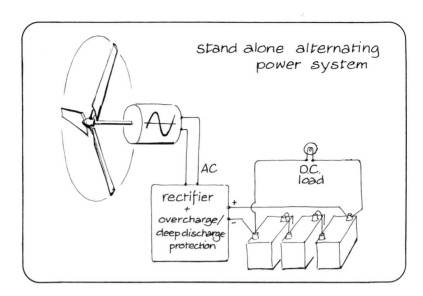

Fig. 3.7   Stand alone alternating power system with rectification
to charge batteries. The direct current d.c. load could be
an invertor to produce alternating current a.c.

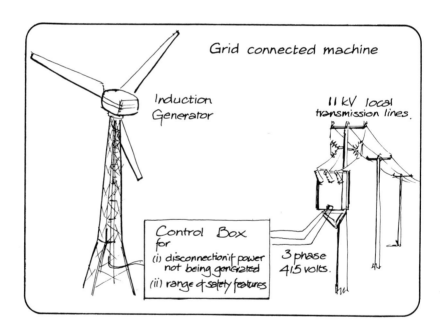

Fig.3.8   Grid connected a.c. turbine. The grid controls the turbine
rotational speed within the necessary limits.

There are many other possibilities for stand alone systems with innovative forms of generation and control. Of particular importance are systems with load management control – see Chapter 9, case studies 2 and 4. In this type of system, the electrical load is automatically adjusted to maintain the turbine frequency constant in changing wind conditions. The advantage is that the use of the power matches what the turbine can give, and the control mechanism is relatively cheap, being electrical rather than mechanical.

3.5.2 Electricity grid connected systems

A most convenient power supply is electricity from a mains grid. For those on a grid, the major expense is the price of the electricity, and so private generation from a wind turbine may be attractive. By connecting the SWECS to the grid supply, the normal electricity circuits can be used (see Figure 3.8). When there is adequate wind, the local supply is used and substitutes for the mains electricity. When there is inadequate wind, power is drawn from the grid in the normal way. Moreover if the local SWECS produces surplus power, the excess can be sold to the grid. Meters are installed to measure separately the imported and exported power. Chapter 5, Section 5.5.3 and Appendix 3 give further details about the regulations for such arrangements.

The usual systems have almost fixed rotor speed operation, whatever the speed of the wind. In this way the frequency of the electrical output through the gearbox and generator is made the same as the mains supply so the connection can be made directly. There are two generator systems for this:

** **Synchronous generation** The frequency of the generator output is entirely fixed by the turbine rotational frequency through the gearbox. Thus the output frequency is in synchronism with the generator shaft speed. Normally these systems can generate electricity even if the mains power fails, so fail safe cut-out equipment has to be installed to prevent electricity going into the grid and harming maintenance and repair workers.

** **Asynchronous generation** The frequency of the generator output is controlled by the excitation from the main supply. Consequently the turbine rotation speed can vary slightly and is not in exact synchronism through the generator with the grid. The normal generator for this is an induction generator with the magnetic excitation drawn from the grid. Note that an induction generator can also generate in a stand-alone situation if self-excited by capacitors.

The great advantage of induction generators is the cheapness of the system and the expected inability to generate when the mains has failed. Therefore line repair operation elsewhere on the grid should be safe. The disadvantage is that the induction generator disturbs the electricity phase relationship on the line. This is expressed by saying that the induction generator draws reactive power from the grid, and this may be upsetting to other users on the grid line. However it is possible to compensate for this reactive effect by incorporating capacitors on the line at the installation.

Another difficulty with induction generators is the possibility of their working as a motor if the wind speed falls. In this way the wind turbine becomes a fan and not a generator! Cut-out mechanisms are installed to prevent this motor action occurring, although initial motoring is often useful to start a machine in light winds.

Most systems are designed such that the rotor and generator are run up to speed using the wind. When the rotor is at synchronous speed the rotating generator is electrically connected to the grid. In some machines the generator must motor the rotor up to full speed before it is capable of generating power. In these circumstances great care must be taken not to exceed the limits on starting current imposed by the local power utility. Care must also be taken to ensure that the starting torque does not over-stress the rotor system.

If a single phase output is required, instead of the more common three phase output, then a single phase generator should be used. However grid connection is not in practice so easy or efficient. Most grid connected SWECS have three phase generators.

If the aerogenerator is to be grid connected, then metering will have to be installed. This metering will be required to measure the kilowatt hours being delivered to the power utility's grid, and to measure the power being consumed by the aerogenerator auxiliary loads and the operator's local loads. Thus the minimum metering required for a grid connected aerogenerator is a kilowatt hour meter to record exported power delivered to the grid, and a kilowatt hour meter to record the imported power consumed in periods of calm.

In addition, the power utility may almost certainly require metering to record the reactive power being supplied from the grid if the generator is of the induction type.

The power utility tariffs may define the unit price paid for electricity generated in terms of the time of day and season of the year i.e. the utility may pay more for the power if it is generated during a time of peak demand. If this situation exists then metering to record the time of day that the power was generated would have to be installed.

3.5.3 Grid connected, variable speed systems

The object of operating in the variable speed mode is to allow the rotor to operate at a fixed tip-speed ratio thus optimising rotor efficiency, see Section 3.4.1 and equation 3.3. The electrical system is much more complex than in the fixed speed systems of Section 3.5.2 above. For these reasons variable speed, grid connected, turbines are not common.

One type of system utilises a pole-change induction generator which may have up to a maximum of three synchronous speeds. The operating speed of the generator can be selected according to the average wind speed. The gains offered by this switching ability have to be compared with the increased cost of the generator and the increased complexity of the control system.

Other systems use a d.c. link between the rectifier and the invertor. Thus the variable voltage and frequency output of the generator is converted to a fixed voltage and frequency of the grid.

Many other systems of generators and controls are possible that should give increased efficiency over simple systems. Usually such systems are more expensive to buy, and so few have been deployed for commercial purposes.

3.5.4 Switchgear
The purpose of switchgear on a grid connected machine is to protect and isolate the generator from the grid should a fault occur, and to provide electrical isolation for maintenance.

In the former case the power utility is concerned that the aerogenerator can be quickly and safely isolated from the grid should a fault develop on it. Likewise if a fault occurs on the grid, then the aerogenerator must cease transmitting to the grid.

When the power utility is carrying out maintenance on cables and overhead lines it must ensure that the line cannot become energised and so is safe to work on. The utility would thus require an isolator switch which can be locked in the open (isolated) position at the point of connection of the aerogenerator to the grid.

3.5.5 Control and maintenance
The degree of complexity of the control system is dictated by the intended use of the aerogenerator. For small machines of tens of kW capacity start/stop controls dependent on wind speed are normally sufficient. Larger machines will have electronic control systems.

For grid connected machines of capacity about 100 kW, a computer will usually control the operation of the movable parts of the blade, the connection of the generator to the grid, and monitor the health of the aerogenerator. Care has to be taken to ensure that the computer is protected against earth loops, lightning, electromagnetic noise and voltage spikes. The control programme should be such that it is readily understood and easily altered if required. A modular system is preferred for this reason.

4   VERY SMALL MACHINES

Contents

4.1   INTRODUCTION

Wind turbine rotors range in diameter from about 30 cm to about 100 m, i.e. from the very small to the very large.   In this Chapter we shall concentrate on the very small turbines, which have distinctive applications and are usually for battery charging.   Their power generating capacity is usually below 500 W.   However their purpose is not strictly for power production as such, but to provide low power of high value, such as for lighting or communications.

Manufacturers of large production machines are always surprised to realise the extremely large numbers of these very small turbines.   For instance one U.K. company now manufactures each year thousands of 50 cm diameter turbine generators with an annual turnover of about £200,000. The systems are sold for battery charging for yachts, caravans, farm buildings and remote locations.   Even if the benefit to the user is only

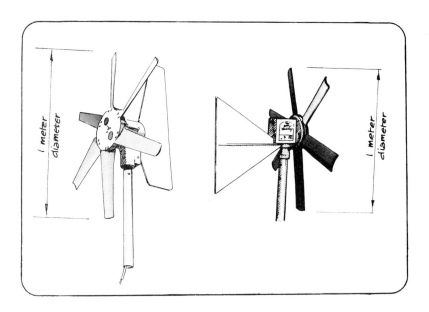

Fig. 4.1   Small wind turbines for battery charging

There is no need to convince people who live with diesel generators of the added advantages of not having the noise of a nearby combustion engine.

The theory of producing electricity by rotating magnetism is beyond this book, but basically the requirement is for a rotating shaft which spins magnets past stationary coils of wire (or vice versa) and so produces electricity. The simplest type of electrical generator for small turbines will consist of spinning permanent magnets passing a static set of copper coils and inducing electricity into them. This system is widely used, but it has two main drawbacks; the control of power is only by control of the current and not the magnetism, and permanent magnetism will attempt to attract the ironwork of the generator thus making the generator difficult to start. This effect (known as 'cogging') is best demonstrated on the bicycle dyno-hub generator, and can lead to significant loss of efficiency when such generators are used with a small wind turbine.

More recent developments have led to 'iron-free' or air-gap magnet alternators that have no iron in them apart from magnets, and thus have no iron to create cogging. Also the effect can be virtually eliminated by careful design of the magnetic field within the generator. Coupled with developments in power electronics, permanent magnet alternators of small sizes are now competing with the more conventional types of generator where the magnetism is produced electrically by additional coils as described below.

Generators without permanent magnets use internal electric coils which themselves produce magnetism when energised by a separate current. Various methods, including external sources of electricity, are used to produce this current. Such a 'wound-field' means that the amount of magnetism may be controlled, and hence the electricity output is also controlled. The disadvantage is that this control current must be fed to coils which are rotating on the shaft, and hence electrical contact brushgear is needed. A motor car alternator is a generator of this type, but is designed for extremely high rotational speeds and would normally be useless for wind turbine operation. Certain low speed wound-field alternators do exist for certain specific markets, but these must be treated with caution as their lifetime may not be long enough for the many years of wind turbine use. Other types of more complicated electrical generators exist, which mainly have the aim of eliminating the brushgear.

4.5 BATTERIES

It is useful to understand some basic terms used to describe batteries.

Volt (V). This unit represents the electrical potential (or pressure) of a system. A single lead-acid cell has a nominal voltage of 2 volts. A single nickel-cadmium cell has a nominal voltage of 1.2 volts. Thus a 12 volt battery would require either 6 lead-acid cells or 10 nickel-cadmium cells wired in series.

Ampere (A). This is the unit of electrical current which is dependent on the voltage of a system and the resistance (R) of that system.

Ampere hours (Ah). This is the measure of the capacity of a battery. It is usually expressed for a specific time interval of discharge. For instance, a battery rated at 170 Ah will, from fully charged, deliver 34 Amps over a 5 hour period before it is fully discharged. (At lower current levels, the discharge time and capacity is effectively increased and vice versa.). It is generally undesirable to recharge a battery too rapidly. As a rule of thumb, a battery should be charged over a period of about 10 hours.

Watt hours (Wh) This is a measure of electrical energy. A watt (W) is the product of the voltage and current in a system.

(Watts = Volts x amps). Thus a 12 volt battery rated at 170 Ah can store 12 x 170 Wh = 2040 Wh (or just over 2 kWh).

The two most common types of battery available for storing electricity are those based on the lead-acid cell and those based on the nickel-cadmium cell. The former is extremely acid and the latter is extremely alkaline - so do not mix equipment.

Nickel-cadmium cells are available in a wide range of power ratings, but have a voltage of only 1.2 V per cell, and are more expensive than lead-acid types. The great advantage of the nickel-cadmium cell is its ability to stand high rates of discharge and long periods of total discharge without damage. Users considering nickel-cadmium cells will normally be considering an expensive investment, and should carefully discuss their proposed application with the wind turbine's supplier.

The vast majority of small wind electric systems operate using the conventional **lead-acid cell.** This comes in a wide range of forms and the initial assumption is that automotive batteries will be acceptable. Whilst this is the case for some very small wind systems, in the vast majority of applications automotive batteries are not acceptable because of their short life span. These batteries are designed for the high current drains of starter motors, and the frequent charge-discharge cycling nature of a wind turbine system causes early plate damage. The type of battery most suitable for aerogenerators is as developed for traction work, e.g. milk-floats or golf carts. These cells are available in either flat plate or tubular plate design (the latter being marginally better, but considerably more expensive) and come as 2 V, 6 V or 12 V battery banks. To provide more storage capacity these can be linked in series or parallel according to the voltage required.

Note that increasing the voltage by linking d.c. cells in series can lead to dangerous installations, because extremely high currents can be produced, and so safety must be checked carefully.

In general batteries need care and attention to maintain fluid levels and prevent corrosion. However sealed 'maintenance free' batteries are increasingly becoming available and may be acceptable. The advice of the supplier should be studied in detail with regard to siting and the safety aspects of housing large battery systems.

Unless a standby generator is included in the overall system, it is necessary to allow for periods of low or no energy input from the wind turbine. Only detailed investigation of wind records will give a satisfactory indication of the extent of such periods. Generally the user should install battery storage for at least five windless days. So if the daily requirement is 1.2 kWh, the storage capacity should be at least 6 kWh (e.g. 500 Ah at 12 V).

The cost of batteries is obviously an important consideration. In general higher price gives better value. The batteries most suitable for renewable energy systems are described below (1986 prices):

	Approximate capacity in £ per kWh	Number of charge/ discharge cycles
Lead-acid		
Standard	55 – 90	700 – 800
Traction	100	1500
Nickel-cadmium		
Low/medium performance	350 – 440	2500

Lead acid batteries are severely damaged by excessive discharge (deep discharge) or overcharge. The most accurate method of assessing the state of charge is by taking specific gravity readings with a hydrometer. Usually a fully charged battery will have a specific gravity of 1.270 to 1.280. At 80% discharge the reading would be approximately 1.180. Some batteries have floats or other charge indicators built into their structure. This is an extremely useful facility, especially if the device is placed for easy viewing.

With either lead-acid or nickel-cadmium cells, hydrogen gas is given off on charge. This is explosive when mixed with air and very great care should be taken not to allow any naked lights or sparks in the vicinity. Obviously therefore batteries must be sited to have good ventilation. The electrolyte of both types is corrosive and any spillage must be well flushed with water, particularly from skin or clothing.

Electrolyte levels should be checked regularly and maintained as specified by the manufacturer by topping up with distilled or de-ionised water. Terminals should be smeared with petroleum jelly (e.g. vaseline) to prevent corrosion and poor electrical contact. All surfaces should be kept clean and dry.

Finally great care must be taken against electrical short circuits or accidental contact by people. Batteries can produce extremely high and continuous currents that easily produce overheated cables and fires. Batteries linked in series can produce dangerous voltages and physical contact can produce lethal currents. Keep all installations well out of the reach of children and animals.

4.6 SYSTEM CONTROLLERS

A wind turbine generator produces very variable amounts of power and usually generates for less than 50% of the time. Other forms of generation (e.g. photovoltaic, hydro, standby diesel) can be combined to produce a more reliable total system. As the wind aerogenerator changes its output, there is also the opportunity to change the load to maintain a match of supply and demand. All such systems need careful control, and fortunately this is possible with modern components. Often control is possible by relays, but increasingly micro-electronic control is becoming available.

The degree of sophistication in the controller varies widely with the application, the importance of reliability and the availability of finance. The simplest wind turbines are d.c. generators capable of being connected directly to a 12 or 24 V battery pack. With occasional maintenance they require no further controlling. However, as a basic minimum, it is worth considering a system that will prevent the batteries from overcharging (gassing) whilst ensuring that the wind turbine remains loaded electrically. Whether this surplus electrical energy is used, generally as heat, or simply wasted to atmosphere, depends on whether there is a use for it. For example diesel engines can be equipped with dump heaters, or the heat can be dumped into water supplies. The best controllers also prevent deep discharge of the batteries.

In general the owner of any size of wind power system should carefully consider the use of modern load control equipment whatever the size of installation. Control is increasingly becoming cheaper and more reliable than most other components in the system, and so will quickly repay investment.

50

4.7 MAINTENANCE

A small wind turbine will be expected to rotate several million times during its lifetime and will inevitably need some maintenance. Blades that have a good surface and a suitable aerofoil shape will suffer least wear. Blade life is mainly limited by wear due to erosion, although poorly made wooden blades can suffer from imbalance due to moisture absorption.

For many-bladed wind turbines it is worth ensuring that a single spare blade can be fitted, not necessarily a whole matched set. It is worth buying a spare blade if the wind turbine is an essential power producer, and if the back-up service of the supplier is at all doubtful.

Other items requiring maintenance on the wind turbine will vary depending on the detailed design. Electrical brushgear will need replacing at some time. Check if replacing brushes is an easy operation or if experts have to be called.

Manufacturers' recommendations about greasing and oiling vary widely. Some manufacturers supply totally greased-for-life systems that require no maintenance. Others include several grease nipples on the wind turbine for regular lubrication. This also ensures that the user will inspect the machinery once a year. Such inspection cannot be over-emphasised in importance. Every user should examine their wind turbines in detail on a regular basis, and, if possible, keep a log book as a record.

5   REGULATIONS AND INSTITUTIONAL SUPPORT

Contents

5.1   INTRODUCTION

Wind turbines are fairly large pieces of equipment that must stand outside in an open environment. It is therefore only reasonable that there should be regulations for their safety and appearance, and that their performance should not unjustly interfere with other people. Many of these regulations are also to the benefit of the owner because a better standard of equipment is available on the market. This chapter has been written in a legalistic manner to cover the many regulations that could apply. It is indeed a formidable list and may appear discouraging to you as a potential owner. However in practice, as with all legal matters, it is not as bad as it seems, especially if you seek the advice of the authorities personally. You are also strongly advised to link with others of similar interest by joining a group such as the British Wind Energy Association.

As with all moving machinery and electricity generating plant, small wind energy conversion systems (SWECS) are subject to regulations and statutory requirements. The design and operation of the SWECS should be in accordance with British Standards and Codes of Practice produced by the British Standards Institution or comparable international standards. Standards specifications are also being produced by national accrediting centres such as the U.K. National Wind Turbine Centre of the National Engineering Laboratory, East Kilbride, near Glasgow.

Local planning permission may be required for the turbine structure and any associated buildings. For the owners and operators of machines, insurance cover is advisable both for the plant and for any liability arising from the installation of the SWECS. For grid connected machines cooperation is needed with the local Electricity Board.

Although this may appear to be a formidable list of actions, the knowledge and expertise of the professional staff of statutory bodies and organisations dealing with these matters are readily available to assist the prospective SWECS operator. In addition, professional advice can be provided by consultants specialising in this work.

This chapter is not intended to be an authoritative interpretation of the legal and other requirements, but it is intended to help prospective SWECS operators in such matters as local planning, environmental issues, statutory requirements, discussions with Electricity Boards, insurance and local rates. Some useful addresses are given at the end of the book, Appendix 2. The chapter is specifically written with regard to legislation and planning in the U.K.

5.2 LOCAL PLANNING

The installation of a wind turbine, or a group of turbines called a wind farm, Fig.5.1, must satisfy the requirements of town and country planning. This is because the installation will come within the meaning of 'development' as defined in the town and country Planning Act 1971. Planning permission will normally be required from the local planning authority unless it is 'permitted development'.

Fig. 5.1   Wind farm of medium size machines.   Good planning is essential.

For example the erection of new agricultural or forestry buildings is 'development' under the Act, but many of these buildings come within what is known as 'permitted development' under the Town and Country Planning General Development Order 1977. This authorises building and engineering operations for agricultural purposes. Buildings and structures must not exceed 465 square metres in area and the height must not exceed 12 metres, or 3 metres if within 3 kilometres of the perimeter of an aerodrome.

Any permitted development proposed in specified 'areas of natural beauty' must not be started before notification to the local planning authority, who must have fourteen days in which to make any specific requirements on design and external appearance.

Some classes of permitted development, such as the construction of buildings or erections exceeding 20 metres in height, have to be advertised in the local press and a site notice has to be displayed.

The application for planning permission is made on a form obtained from the local planning authority, which will normally be the district council. Prospective SWECS operators are strongly advised to seek the advice and assistance of the professional staff of the planning authority at the earliest possible stage.

If prospective SWECS operators wish to know about planning permission before incurring too much expense, they can apply for 'outline planning permission'. This can be given by the planning authority subject to their later approval on siting, design and external appearance.

The local planning authority must give a decision on valid planning applications within eight weeks, unless the period is extended by mutual agreement.

In addition to making an application for planning permission, the prospective SWECS operator must give notice of intention to carry out work under the Building Regulations on a form obtained from the local Council.

5.3 ENVIRONMENTAL ASPECTS

5.3.1 Visual/aesthetics

One of the most obvious environmental effects of a SWECS is the visual or aesthetic impact. The local planning authority will give careful consideration to this aspect of any planning application. Colour and good design will be important in creating a pleasing reaction from the public and planning authority.

5.3.2 Acoustic Noise

SWECS are unlikely to produce major noise problems and there should be no more difficulty with wind turbine noise than there would be over noise from any small-scale industrial development.

Methods for assessing acceptable noise levels for industrial developments have been developed over many years. They usually involve an evaluation of existing (background) noise levels and require that these should not be increased by more than a certain amount. Guidelines are available of background noise levels which might be expected if it is not possible to make sound measurements, for example in British Standard BS 4142:1967, relating to mixed residential and industrial areas. Acceptable noise levels indoors in rural areas are discussed by the House of Commons Committee on the Problems of Noise (see Bibliography).

Background noise differs with the type of area, for example, urban, rural, recreational, and also with factors such as time of day, human activity and weather. Because of this, noise levels have to be expressed in statistical terms. When new plant is being evaluated, a statistical measure of fairly quiet background conditions is employed as a base line such as the levels which are exceeded 90% or 95% of the time at the location being considered. If noise from any new plant is only allowed to exceed such a level by a few dB(A) (a unit for sound measurement) then it will only be noticeable for a small fraction of the time because noise in the area will usually be higher. Furthermore, experience has shown that on the occasions that it can be heard, the extra noise will not give cause for complaint.

Some noise sources have special tonal characteristics that make them more noticeable than others even though they register the same noise levels. Standard methods can be applied to allow for this in planning the development.

Wind turbines are something of a special case at present and are not covered fully by existing standards, though siting criteria are being developed. The uncertainties are especially relevant to the very large machines designed to operate at high average wind speeds, for example, 10 metres per second. The problem arises because the criteria discussed above are all related to fairly quiet background reference periods. These usually occur when there is little or no wind, since the wind itself generates significant amounts of noise around buildings, trees, shrubs, grass, water surfaces and even around the head of or in the ears of the observer. One method for assessing how well this masks noise from a wind turbine has been suggested by Soderquist (1982) in a Swedish investigation (see Bibliography). This indicated that the national 'still air' background noise level should be corrected to allow for masking at wind speeds greater than about 5 metres per second (measured at 10 metres height), with the correction increasing by about 12 dB(A) for every doubling of wind speed. It is likely that this method will be refined as experience accumulates on wind generated background noise.

Work is also under way on the 'noticeability' of wind turbine noise, because of its special character. The noise has a number of components including gearbox noise, which is controllable by normal sound proofing methods, and blade noise which manifests itself as a rhythmic swishing. The latter cannot be completely eliminated and will ultimately limit how close wind turbines can be located to dwellings. Some very large machines, designed so that the blades are downwind of the tower during operation, generate very low frequency noise. This is due to the rotor blades cutting across the wind flow behind the tower and has led to adverse publicity about wind turbine noise. However this has only been associated with the large, megawatt-scale machines, well above the size of the units being considered in this book. Noise standards usually incorporate ways of allowing for tonal characteristics by correcting allowable levels by a few dB(A). These can be used if necessary for a particular wind turbine but may be overstringent. Research is under way which will help to clarify this uncertainty.

Ways of reducing wind turbine noise are also under continuing study and prospects for siting of turbines are promising. For example, it has been suggested that it should eventually be possible to locate a 300 kW, 25 metre diameter blade machine within 200 metres of permanent dwellings. At present somewhat greater separation distances would be appropriate depending on machine design and location. For the smaller SWECS, the corresponding distance would be much less.

5.3.3 Telecommunications interference

Telecommunications services, such as radio and television broadcasting, or fixed and mobile radio services, can be affected by interference from a SWECS. This can result from the tower or the blades causing obstruction or reflection of the signal, or by electrical interference from the generating equipment. Interference may be more troublesome in areas where radio and television signals are weak. Factors which can affect interference are the frequency and strength of the transmitted signal, the location of the SWECS in relation to the transmitting and receiving stations, and the size and construction materials (especially metal) of the SWECS. Site selection is a critical factor in minimising interference.

Prospective SWECS operators should be satisfied that interference problems are not expected to occur. Expert advice can be obtained from the Radio Investigation Service, Directorate of Radio Technology, Department of Trade and Industry (see list of useful addresses).

5.4 STATUTORY REQUIREMENTS

SWECS have to comply with statutory requirements with respect to their design, construction, operation and maintenance. These are set out in the relevant Acts and subordinate legislation. It is important for the SWECS operator to know what legislation applies in his particular case.

5.4.1 Health and Safety at Work etc. Act 1974

The legislative framework for the protection of the health, safety and welfare of people at work, and of others affected by the activities of people at work, is the Health and Safety at Work etc. Act 1974.

The Act is superimposed over existing health and safety legislation, such as the Factories Act 1961; the Agricultural (Safety, Health and Welfare Provisions) Act 1956; the Offices, Shops and Railway Premises Act 1963 and the Regulations made under them. Such legislation remains in force until replaced by Regulations and Approved Codes of Practice made under the 1974 Act. However, enforcement has been changed by the Act with the establishment of the Health and Safety Executive. Amending Regulations have been made under the Act which have affected the earlier legislation.

The Health and Safety at Work, etc. Act applies to virtually all premises but the previous Acts and Regulations apply only if the SWECS is in a particular category of premises. The Act imposes duties on all 'persons at work', whether employers, employees, self-employed or others (such as designers, manufacturers and suppliers) but with the exception of domestic servants in private households.

Many prospective SWECS operators may already have responsibilities under the Act or previous health and safety legislation, but some may become subject to such legislation for the first time. A guide to the Act and a series of leaflets which will help SWECS operators to understand their responsibilities under the Act can be obtained from Government Bookshops or the area offices of the Health and Safety Executive, which are represented by the Factory Inspectorate and other Inspectorates (see Bibliography). Advice on the provisions of the Act as they relate to a particular SWECS installation can be obtained at these area offices.

5.4.2 Factories shops and offices

The Factories Act 1961 deals with the safety, health and welfare of people employed in a 'factory', which can be taken by a SWECS operator to be any premises in which people are employed in manual labour to make, alter, repair, finish, clean or adapt articles by way of trade or for purposes of gain. A 'factory' need not be a building and can be open air premises. The exact position of electricity generating premises, with regard to their definition as factories under the Factories Act 1961, is contained in Sections 123 and 175 of the Act. Where persons are employed at a SWECS, the premises would almost certainly be classified as a factory for the purposes of the Factories Act 1961.

57

If SWECS operators are in any doubt whether their premises is a 'factory' under the Act, they should obtain advice from the local office of the Health and Safety Executive. There is a legal requirement, in any case, to give the Factory Inspectorate a month's notice before starting up a factory. Special Regulations, such as the Electricity (Factories Act) Special Regulations 1908 and 1944 have been made to deal with special conditions (see Section 5.5.1).

Factory Act legislation is enforced by the Factory Inspectorate which is now an Inspectorate of the Health and Safety Executive.

The Offices, Shops and Railway Premises Act 1963 is modelled on the Factory Act and its scope covers the premises indicated in its title. It is not likely to be of general interest to SWECS operators but may have some relevance to a particular installation.

5.4.3 Agricultural premises
The Agricultural (Safety, Health and Welfare Provisions) Act 1956 and the Regulations made under it deal with safety, health and welfare of people working in agriculture. Details of amending Regulations made under the Health and Safety at Work etc. Act and other useful information are given in a Guide to Agricultural Legislation by the Health and Safety Executive (see Bibliography).

5.4.4 Fire precautions
In general, fire precautions are governed by the Fire Precautions Act 1971 and the Regulations made under it. As a result of the Health and Safety at Work Act, amending Regulations have been made in relation to the 1971 Act and to the fire precautions provisions in the Factories Act and the Offices, Shops and Railway Premises Act.

Help and advice on the provisions of these Regulations can be obtained from the local Fire Service, who have the responsibility for enforcing the legislation.

5.5 ELECTRICAL ASPECTS

5.5.1 Legislation
The Electricity (Factories Act) Special Regulations 1908 and 1944 is the general title of the legislation concerned with the generation, transformation, distribution and use of electrical energy in premises to which the Factories Act 1961 applies (see Bibliography). Depending on circumstances, they may apply to a SWECS installation. They are a good guide to compliance with the Health and Safety at Work etc. Act 1974, whether or not these Special Regulations apply to a specific SWECS installation. Note however that the UK Health and Safety Executive plan to replace the "Electricity (Factories Act) Special Regulations 1908 and 1944" by new legislation, the "Electricity at Work Regulations".

The Electricity Supply Regulations 1937 contain provisions regarding the electric lines and works of statutory electricity undertakers (i.e

Electricity Boards) and the supply of energy to consumers and consumers' installations (see Bibliography). The Energy Act 1983 (see Section 5.5.3 and Appendix 3) amends the law relating to the generation and supply of electricity by persons other than Electricity Boards. Regulations are made under the Energy Act to supersede the Electricity Supply regulations 1937 to take account of the provisions of the Act. The provisions of the 1937 Regulations concerning the supply of electricity to consumers and consumers' installations (Regulations 22 to 35) are taken to apply to private generators such as SWECS operators, unless the new Regulations supersede them.

The electrical installation for a SWECS should be wired in accordance with the latest edition of the Institution of Electrical Engineers Wiring Regulations (Regulations for Electrical Installations). Installations which comply with these Regulations are deemed to fulfil the relevant requirements of the Electricity Supply Regulations. The design and installation of electrical equipment needs specialist knowledge, and so professional guidance should always be obtained.

The Health and Safety at Work etc. Act requires an employer to give information to others about the way in which he conducts his activities, where these others, not being his own employees, might be exposed to risk to health or safety in consequence of the way in which his activities are undertaken. It is therefore in the interests of potential SWECS operators to inform the local Electricity Board of the generating plant they intend to install, whether or not it is to be connected to the public network, since fortuitous connection from the SWECS could put other people at risk.

SWECS operators may also need to consider local authority bye-laws on the generation of electricity and any conditions of private and industrial insurance.

5.5.2   Interconnection with Electricity Board networks
Many private generators operate in complete isolation from the local Electricity Board network but there are many instances when SWECS operators will wish to be interconnected with the public supply, Fig.5.2. In this case, they should contact the local Electricity Board as soon as possible.

The ability of a local network to accept the connection of a SWECS will depend on the electrical characteristics of the network and of the machine to be connected. The electricity network can be very different between Boards and even between different parts of the same Board. For this reason, it is essential that the Electricity Board should be given the opportunity at an early stage to consider the suitability of the local network to accept the SWECS. Reinforcement of this network may be necessary and the potential SWECS operator could be faced with the cost of the work, which may amount to several thousands of pounds. A particular example of this is in rural areas when the local electricity network is single-phase only. The SWECS operator might have to meet the cost of providing a three-phase supply to the SWECS if the Board will not accept a single phase arrangement.

59

To assist private generators of electricity, the Electricity Boards have produced two advisory documents, which can be obtained from the Electricity Council (see Bibliographay). One document gives general guidance on the requirements of Electricity Boards for the possible connection of private generating equipment to public electricity networks, and the other gives guidance on the safety implications of the Energy Act 1983.

5.5.3 The Energy Act 1983

The Energy Act 1983 has significantly changed the position of private generation in the United Kingdom. Its broad aim is to encourage the private generation of electricity. An informal guide to the private electricity supply provisions of the Act is available from the Department of Energy or the Scottish Economic Planning Department (see Bibliography). Appendix 3 gives more details.

The Act removes the statutory restriction which prevented the supply of electricty as a main activity of a business by anyone other than an Electricity Board. It also enables a 'private generator or supplier' (defined as a person, other than an Electricity Board or Local Authority, who generates or supplies electricity) to request an Electricity Board to supply electricity to him for his own use or for his customers, to purchase electricity generated by him or to use the Board's transmission and distribution system to give a supply of electricity to any premises. The Board is obliged to make an offer to comply with the request, unless on technical grounds it would not be reasonably practicable to do so.

If SWECS operators wish to become private generators or suppliers, they will be asked by the local Electricity Board to provide technical information about their plant. The information which is required is set out in The Electricity (Private Generating Stations and Requests by Private Generators and Suppliers) Regulations 1984, Statutory Instrument 1984 No.136.

Electricity Boards must establish and publish tariffs for supplying electricity to a private generator or supplier, for the purchase of electricity produced by the private generator and for charges for the use of its transmission and distribution system.

The Act makes it clear that another person receiving electricity from a private generator via a Board's network will become the private supplier's customer and no longer a consumer of the Board. The private generator or supplier may remain a consumer of a Board, and may then ask the Board to supply electricity to premises from which electricity is supplied to this consumer's own customers. This allows the Board's electricity to be a private standby supply when the private generating equipment is unavailable.

The Act provides for disputes between private generators and suppliers and Electricity Boards.   The procedures for determining disputes are set out in the Electricity (Conduct of Proceedings for the Determination of Disputes) Regulations 1984, Statutory Instrument no.135.

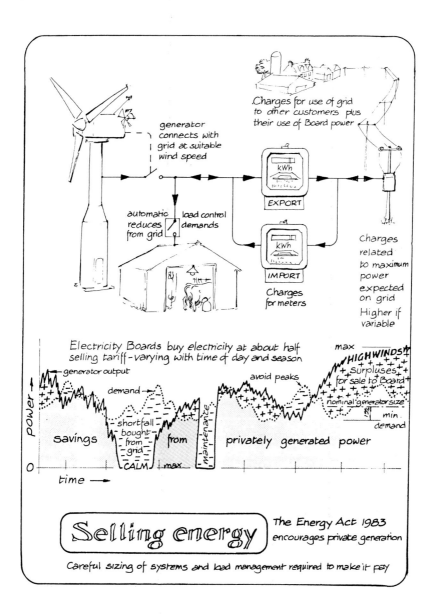

Fig.5.2   Buying and selling electricity by means of the 1983 Energy Act.

5.5.4 Lightning protection

British Standard Code of Practice CP 326: 1965 'The Protection of Structures against Lightning' provides authoritative guidance of a general nature on the principles and practices which experience has shown to be of major importance in the protection of structures against lightning and on the problems of when to protect and when not to protect.

5.6 INSURANCE

As a SWECS is a substantial capital investment, the owner should insure the plant against damage and against any effect which the equipment might have on third parties, such as the general public or neighbours. If the SWECS owner employs people to carry out his business, he will come within the terms of the Employer's Liability (Compulsory Insurance) Act 1969, which says that he must provide insurance to cover his liability for death or injury to employees sustained in the course of their employment.

However, he may already have fire and special perils cover in respect of buildings and other classes of machinery. Such policies can be extended to include the SWECS installation. This would then give cover for damage caused by fire, lightning, explosion, storm, tempest or flood, civil commotion and malicious damage; although the tendency nowadays is for insurance companies to provide cover under an 'all risks' policy for accidental damage.

Experience from Denmark where there are now over a thousand SWECS shows that annual insurance premiums are in the range of 0.5%-1% of the capital cost of the machine.

Engineering insurance should be arranged for the SWECS installation, including the wind turbine, electricity generator, electrical system and tower. The cover can be on all-embracing basis of sudden and unforeseen damage which will embrace "Breakdown and Damage from Accidental Extraneous Causes", such as "Impact or Collision and Collapse", or alternatively, specific defined parts of the cover are available. The SWECS owner may also wish to cover his loss of profit due to fall in takings because of inability to generate electricity from the SWECS because of damage. If the owner can buy electricity from the Electricity Board at increased cost to satisfy his own customers' demand, then these additional costs are insurable.

Under the Factories Act, certain plant in the factory, i.e. lifting machines, air receivers and boilers, are required to be examined at regular intervals by a competent person. The insurance companies can usually arrange for this inspection to be carried out. Although not a statutory requirement, insurance companies are likely to require that SWECS towers and safety devices are inspected by their engineers at regular intervals.

Damage to surrounding property belonging to the SWECS owner can be covered by an extension to an engineering policy for an additional premium. However it might be more convenient to extend an existing fire and special perils policy to include such cover, unless it is already provided for. A general public liability policy could cover third party risks, such as damage to other people's property, but it would need to be specially mentioned to the third party insurer.

As SWECS are relatively new to insurance companies, they will require detailed information of the installation in order to make an evaluation of the risk and determine premiums. Premiums for engineering insurance will vary with the value of the installation, the location of proposed SWECS and the company's valuation of the risk.

Engineering insurers involved in this type of business are:

> British Engine Insurance Limited
> Eagle Star Group Engineering Insurance Limited
> National Vulcan Engineering Insurance Group Limited
> Scottish Boiler and General Insurance Group Limited

The addresses of these companies are given in the list of useful addresses, Appendix 2.

Insurance cover for damage, fire etc. would be carried out by the local offices of insurance companies who deal with the normal range of insurance cover.

The British Insurance Association is the trade association and would be able to give general advice on insurance matters.

5.7 RATES FOR LOCAL TAXATION

5.7.1 Rating assessment
The occupier of any land is liable to be rated and the annual liability to rates is calculated by applying the rate determined annually by the rating authority to the rateable value of the property (or 'hereditament' as it is termed), entered in the Valuation List.

Theoretically, rating valuation is the annual rent which might reasonably be charged for a 'hereditament'. In the absence of direct rental evidence, the British Wind Energy Association has been advised that a SWECS would be valued on a 'Contractor's Test' basis of valuation. This generally uses a decapitalisation rate of 5% on the assumed capital value of the SWECS at the date of the Valuation List (1973 in England and Wales, 1976 in Scotland).

Legal opinion received by the British Wind Energy Association clearly indicates that the land occupied by and in conjunction with the SWECS, together with the structure which supports it, would fall to be included in the assessment. It appeared likely that the plant and machinery of a SWECS would fall to be assessed, 'up to and including .... in the case of electrical power, the first transformer in any circuit, or where the first transformer precedes any distribution board or, if there is no transformer, the first distribution board'. This opinion relates to England and Wales but it is understood that wind turbines would be rateable to a substantially similar extent in Scotland.

The generating plant of the Electricity Boards is rated in a different way that greatly reduces the amount paid per unit of electricity supplied. The British Wind Energy Association is making representations to the Government on the rating assessment of private wind turbines, with the aim of reducing the amount to be paid.

5.7.2 Agricultural land and buildings
The General Rate Act 1967 states that 'no agricultural land or agricultural buildings shall be liable to be rated or included in any valuation list or any rate'. A SWECS used only to supply electrical power to agricultural buildings and land would probably be exempt from rates, as for instance with the SWECS on Lundy Island. However, the position of a SWECS which exports some of its electrical output to the public network is uncertain at the present time.

5.8 GRANTS
The only grants available in the UK that explicity mention wind turbine generators are those operated by the Ministry of Agriculture, Food and Fisheries (Department of Agriculture and Fisheries for Scotland). The maximum capital grant is either 30% or 15%, dependent on the designation of the area under EEC criteria. The grants are available within a complex formula, and are only awarded for certain types of farming. Farmers are strongly advised to seek such grants, but the benefits will depend much on local circumstances.

5.9 SUMMARY

Matters that potential SWECS owners will need to consider are listed below, together with the organisations which could provide guidance and advice. A bibliography is included in the Appendices.

Building regulations	Local council
Design and operation	National Wind Turbine Centre British Standards Institution
Dispute with Electricity Board	Department of Energy
Electrical installation	Institution of Electrical Engineers Local Electricity Board
Electricity regulations and Health and Safety	Health and Safety Executive Factory Inspectorate Electricity Council
Electricity tariffs	Local Electricity Board
Fire precautions	Local fire service
Grants (Agriculture)	Ministry of Agriculture, Food and Fisheries Department of Agriculture and Fisheries (Scotland)
Grants (Industrial Development)	Department of Trade & Industry
Insurance	British Insurance Association Engineering insurance companies
Noise	British Standards Institution Local council
Rates	Local council British Wind Energy Association
Telecommunications interference	Department of Trade and Industry
Town and country planning	Local planning authority

6   BUYING AND FINANCING A MACHINE

Contents

6.1   INTRODUCTION

Power generated from wind turbines can be cheaper than from conventional sources under a variety of circumstances. Considering the capital and maintenance costs of operating a small wind energy conversion system (SWECS) over its working life, the plant may supply energy cheaper than conventional sources of energy such as oil. The economic calculations are based on four main factors:

**   The performance characteristics and capital cost of the SWECS.

**   The annual average wind speed at the proposed installation site, and, in certain circumstances, the way the wind varies throughout the year.

**   The cost of the conventional energy at that site, including supply and transmission costs (e.g. electricity, oil, gas etc.).

**   The extent to which the SWECS output is usefully used, given the variable nature of the wind.

**   The ability of a SWECS to survive for at least 15-20 years within its maintenance budget, and to generate at its designed rate throughout this lifetime.

The technical requirements for small machines with rated powers to about 150 kW are now well understood. Consequently commercial SWECS should operate to specification and survive with regular maintenance for at least 15-20 years or more. However not all SWECS presently manufactured live up to their technical specifications regarding production of energy and lifetime. Nevertheless there are many purchasers who purchase machines still under development or that are expensive, despite the questionable economic return on that particular machine. For instance several Electricity Boards have purchased prototype machines in order to gain advance experience for a new service of reliable future energy.

With improvements in what is still a new technology, there will be significant real reductions in the capital cost of SWECS in the next few years.   In addition it is anticipated that the real cost of fuels will rise in the future, so making the economic case for wind machines even more reasonable. Falling production costs for SWECS and rising fuel costs will steadily increase the number of economic wind sites.

6.2   MAJOR FINANCIAL CONSIDERATIONS

The initial investment cost of a SWECS is relatively high, compared with oil fuelled generators. Present costs of installed SWECS equipment are typically £500 to £1000 per kilowatt of generating capacity. Financial institutions that loan money for new construction and plant may be expected to be conservative in backing new technologies, particularly in the present economic climate, so investment in a SWECS may not be considered enthusiastically by them.

Unfortunately, whilst there are a number of arguable reasons for government to encourage the use of new technologies and develop new industries by providing finance, there are currently in the United Kingdom no general subsidies or grants from central government for wind turbine owners. Commercial or agricultural organisations may however be able to obtain assistance as with other developments, see Section 5.8. In general the financing options must be arranged by the purchaser, often with the help of the manufacturer.

The manufacturer may offer assistance by obtaining finance or by arranging leasing facilities. For the purchaser there may be a number of financial options and/or incentives available. These are :

Tax reduction
Whilst private individuals in the UK cannot obtain tax credits for wind turbines as in many European countries and North America, it is hoped by the BWEA that these may become available. However there are possibilities for most agricultural or commercial purchasers to offset costs against tax liabilities. Such tax benefits may effectively reduce the capital costs of SWECS by perhaps 25%.

Grants for Agriculture
Direct grants are now available from the Ministry of Agriculture, Fisheries and Food, (Department of Agriculture and Fisheries in Scotland) to offset the cost of SWECS for certain agricultural users. Up to date information is available at your local MAFF office. Grants of 15% or 30% are in principle available.

Industrial grants
There are a number of possible regional incentives, such as grants, but these will depend very much on the individual case.   Such grants are particularly available for capital investment in industry, and may be about 25% of the capital cost Similarly there are possibilities of obtaining local home improvement grants. For details contact your local

Department of Trade and Industry Office, Regional Authority or Development Area Office.

Selling electricity

Selling energy which is excess to purchasers' needs is obviously an important benefit. With the introduction of the UK Energy Act (see Chapter 5 and Appendix 3). Electricity Boards have an obligation to purchase electricity from, and supply electricity to, private generators at reasonable rates. These rates and connection charges open up a whole new area of economic calculations and decisions on whether or not to have wind turbines connected to the grid.

Initial reactions to the Energy Act with regard to generation of power from the wind were summarised in the 1984 Proceedings of the BWEA annual conference. There are several kinds of financial charges or payments. These are:

> Capacity related charges   The Board will charge the private generator for the maximum power expected to be imported or exported.   This refers to the cost of the Board installing generator plant and transmission lines. These charges will be high per unit of power if the rate of exchange of power is variable.   It is very important to avoid excessive peaks in either imported or exported power.

> Metering charges   The private generator has to pay for special instruments needed to meter the import and export of power in grid connected machines.

> Customers of the private generator   The private generator may use the normal grid lines to supply other customers. The Board will charge for the use of such lines.   Moreover any power used from the Board by such customers is chargeable to the private generator.

> Import and export of power   In general an Electricity Board will buy electricity at about half the cost per unit that it charges for sale.   However both buy and sell rates will vary through the day and the season.   Thus much more is earned by selling power in the daytime during winter, than at night-time in the summer.   In the UK there are substantially different rates offered at different times by the different Boards.

The best value of SWECS generated power is given by substituting for grid purchases. In this way an owner achieves the highest value for the power produced.   Often the system should be sized so that maximum output of the SWECS is slightly (say 15%) greater than the usual minimum demand of the user loads.   Thus normally power will not be exported to the grid and an oversized machine cost has been avoided.   Further improvement can occur when manual, or better still automatic, control of the user loads can be arranged to work in conjunction with the SWECS generation.   In windy conditions loads are switched on (e.g. heaters, water pumps) and in less windy conditions these are progressively switched off so power is not drawn from the grid.

In summary therefore, a user can only expect to gain from selling power to the grid if there is very careful sizing of the system and careful load management.

Stand-alone-operation
The option of not connecting a SWECS to the grid should be considered. This avoids all complications involved in connection with the Electricity Board's network grid, and allows different technical arrangements. In practice such systems are likely to be for providing electric heating in place of using fossil fuels or grid supplied electricity, or providing light and other electricity supplies where no grid supply exists.

Other financial aspects
Other minor, but possibly important financial points include the cost of insurance, the owner's liability for damages, the cost and nature of warranties and guarantees, and the costs and usefulness of service contracts for maintenance. These factors are considered either in the following sections, or in the previous chapter (Chapter 5). Disincentives could be the cost of local taxation rates, although the BWEA hope Government action will lessen this, see Section 5.7.

6.3 BUYING A WIND TURBINE
In principle buying a wind turbine should be no different from buying any item of equipment such as a tractor or a car. Indeed in countries like Australia one can purchase wind turbines, water pumps and battery chargers through mail order catalogues or from local hardware stores. However buying larger electrical generating machines for significant energy production is not yet so easy.

In the near future in the UK there will be certification for machines to give guidance on their suitability for use. We also expect publications by consumer organisations to compare performance and cost. Credit companies may offer tailor-made financing options including leasing. The UK National Wind Turbine Test Centre has been established at the National Engineering Laboratory, East Kilbride, Scotland. This should lead to much more consumer information being available regarding reliable machines. Nevertheless purchasers must always scrutinise manufacturers' sales information and consider the economics of wind turbines.

6.4 ECONOMICS OF SWECS OPERATION

This section considers the general aspects of the economic considerations. Worked examples of certain accountancy procedures are given in Appendix 4.

6.4.1 Introduction

Only in the narrow (but important) case of wind turbines working under well defined conditions, and probably tied into a known and existing utility network, can accurate economic assessment methods be used. The majority of installations in the small and medium categories are unlikely to be amenable to such a strict treatment. More often than not, many factors which are difficult to quantify determine the outcome. Nevertheless economic factors, where they form part of an assessment, should if possible:

**   Be simple, clearly defined and meaningful to the prospective purchaser;

**   Be based upon good wind data from the actual site in question or failing that, from a nearby metereological station;

**   Include the 'extra' costs of erection and commissioning, which may amount to 20% of the factory costs;

**   Include full costs, e.g. batteries, wiring, meters;

**   Use actual fuel bills to compare conventional supplies with the SWECS alternative;

**   Include grants if these are available

The wind resource is the most important feature in establishing suitable locations, so the same wind turbine may be highly economic at one site and grossly uneconomic at another. Other factors can be advantageous e.g. transportation costs of conventional fuel are frequently double or triple the base fuel cost and this may create a favourable competitive market for SWECS. Care should always be taken to establish the actual cost of fuel at the proposed site. One distinct advantage of wind turbines for remote locations is that they can provide energy when conventional energy systems may be out of action due to industrial disputes or adverse weather conditions.

The cost of a wind turbine will be quoted at an 'ex factory' price. In addition the cost of transportation, erection and commissioning will be quoted separately, as it is dependent not only on delivery distance but also on the actual geography of the location. Erection and commissioning are likely to require two visits including foundation preparation.

Annual maintenance costs may be expected to be 2% -4% of the capital cost. Usually such costs fall during the initial period and rise again towards the end of the machine's life. It is essential that spares are available during the whole life of the machine and that subsequent improvements in the design are issued in planned fashion and notified to previous purchasers. For general maintenance most manufacturers are offering, at the least, annual lubrication and a 20 - 25 years' wind turbine active life. Materials and stress levels are chosen with specifications such as these in mind.

6.4.2 Depreciation

Depreciation is an accounting concept used to describe the loss in value of an asset that occurs through usage and the passage of time. This can come about as a result of the fall in price of the new asset, by physical deterioration, and through a fall in value of the asset as more cost effective machines become available.

Depreciation represents that part of the cost of an asset to its owner which is not recoverable by him when the asset is finally put out of use. It follows that provision against this loss of capital is a cost of conducting the business and is not dependent upon the amount of profit earned.

When the rate of depreciation to be charged on the wind turbine has to be decided, the accountant can be in some difficulty, for not all components of the turbine will decline at the same rate. The historic cost of the wind turbine will include the cost of the machine and the cost of the tower at the factory gates, the cost of getting them to the chosen site, the cost of foundations and erection, and the cost of professional services associated with this activity.

As depreciation should represent the cost to the business of the annual loss in asset value, it should be known how long the turbine is likely to last. It is reasonable to assume that a professionally manufactured wind turbine, with regular maintenance and repair will operate for at least 20 years. The tower, foundations, delivery and fees (the unlifed parts) can be depreciated at the appropriate rate of say 5%. The rotor, the hub, and other items of the turbine mechanism may have a fixed period of hours run or power produced before they are removed for replacement (lifed parts). If this is the case, the manufacturer will advise the customer of components' lives and replacement costs. The first costs of these items are deducted from the total cost of the turbine and will have to be set against the hours run or the kilowatt hours generated in the accounting period.

Depreciation can be charged as a constant proportion of a decreasing balance of the fixed asset value (the decreasing balance method), or as a fixed amount deducted from that capital sum each year (the straight line method).

The first method has the advantage that the depreciation charge in the first years of use is high when the cost of maintenance and parts replacement is low, thus giving a more uniform operating cost. The depreciation charge never quite reaches zero no matter how long the machine is used. The disadvantage is that the unevenness of the loss in

71

market value of the wind turbine over several years is not represented, but it does better reflect the diminishing losses in value that occur in later years as the turbine approaches the end of its useful life.

The straight line method has the advantage that the equipment is written off completely in a finite time. However the residual value upon which the depreciation rate depends is very difficult to establish. It is more difficult to decide when replacement should take place after the machine is paid for, and it may be sold before its time.

6.4.3 Insurance cover (see also section 5.6)

Insurance is necessary for the wind turbine, not only to protect the owner from the consequences of third party claims arising from premature failure of the machine – an unlikely event – but also to provide employer liability cover for any employee working on the turbine, as well as the more usual risks of loss of some or all of the equipment by fire, storm and tempest. There may be additional maintenance contract costs as well. When wind turbines are in wide use, loss of use insurance will become available. In the absence of field data, the insurance costs are assumed here to be 2.5% of first cost. However insurance rates in Denmark and the USA where there are many thousands of machines, is usually less than this.

6.4.4 Local taxation of wind turbines (local rates)

The legal background to tax assessment for local rates is given in section 5.7. In theory, the rating valuation is an annual rent that could be charged. Wind turbines, being a new development, have no rental evidence and so rates will have to be assessed on a 'Contractor's Test' basis of valuation. The BWEA sought the advice of Chartered Surveyors, J R Eve and Company who told them that the 'Contractor's Test' in England uses a decapitalisation rate of 5% on the assumed capital value of the wind turbine at the date of the valuation list. This is the 'tone' date and operates under Section 20 of the General Rating Act 1967, and is fixed at 1973 in England and 1976 in Scotland. Valuations at later dates are adjusted by a factor to bring them to the same basis of cost had they been assessed on the date of the valuation in 1973 or 1976. If the wind turbine supplies power to an agricultural establishment, then rates are not charged.

At the time of publication of this book, various test cases are being made against the payment of local rates for wind turbines. The BWEA is closely following events.

6.4.5 Cost of maintenance

The costs of operation of a wind turbine are not often anticipated correctly, for they are influenced by the location in which the turbine operates. The prospective purchaser is advised to get contractual undertakings on expected operation and maintenance (O & M) costs from the manufacturer as part of the purchase negotiations. Wind is a new form of power source and unless the characteristics of the wind at the chosen location are well defined and understood, the required power put out by the wind turbine will not be achieved.

The manufacturer should have had sufficient experience to give accurate O & M cost figures for the equipment.

Operation and maintenance costs are very strongly influenced by the degree of sympathetic understanding that the machine receives from those it serves. In-service experience of some island communities has shown that the wind turbine brings with it, through the provision of ample low cost power, a higher level of life-style. In such cases the wind turbine receives a close and informed support from the community. A potential purchaser is advised to look into these aspects of the investment with care.

6.4.6 Cost of energy displaced

The cost of the power put out by wind turbines and that purchased from the grid may not be compared directly. Wind turbines installed for purposes other than battery charging often produce more power than expected. It is advisable to design a system for a surplus of power because, once installed, such systems have more and more demands placed upon them because they are there. In a domestic application, the wind turbine may heat the water for washing, and provide for house heating where solid fuel had been used. All such additional benefits must be included in the evaluation of the new investment.

In residential and industrial installations, it is almost certainly more economic for the operator to use his own wind electricity than to feed power to the grid and buy grid electricity at another time. The selling price of electricity is usually substantially lower than the average buying price.

Diesel electric generators are widely used in farming and in telecommunications for remote power supplies. The cost of fuel for a kilowatt of electricity is high when transportation charges and diesel generator efficiency are brought into the calculation. In such installations, electricity costs of about 15p per kilowatt hour are not uncommon, and abroad in island communities, costs of 75p per kilowatt hour have been reported.

6.5 SUMMARY OF ECONOMIC FACTORS

The following data will be needed for financial purposes. These are explained more fully in Appendix 4.

> System size (kW)
> Cost per ex-works
> Cost of delivery
> Installation cost
> Professional fees
> Depreciation rate on fixed asset
> Fixed asset proportion of installed cost
> Grant proportion on fixed asset
> Grant value
> Annual operational and maintenance costs
> Land rates

Property rates
Insurance premiums
Borrowing ratio (borrowing/total investment)
Borrowing interest rate
Borrowing period
Alternative fuel costs
Fuel price increase rate
Net Present Value depreciation factor
Rate of Value Added Tax

6.6   CONCLUSIONS

In this chapter, and in the associated Appendix 4, an attempt has been made to show how the costs of wind turbine operation are assessed, and how these costs are effected by inflation and depreciation.

It cannot be over-emphasised that the quality and serviceability of a wind turbine are most important, for a wind turbine must work continuously for its economic life. Money spent in investigating the market and wind conditions at the proposed site is never wasted, for it returns with interest in lower operating costs and more power generated.

Perhaps more than most machines, wind turbines require to be evaluated by considering the costs of purchase, operation and maintenance throughout their operating life. Twenty years is a long time for any machine to work, and the total quantity of the power generated may be more than anticipated. Regular, thorough, and skilled maintenance will be the best safeguard against premature failure and so make it possible for the expectations of the investment to be realised.

7   INSTALLING AND OPERATING A SYSTEM

Contents

7.1   INTRODUCTION

The previous chapters have given the background detail for deciding whether or not to install a wind energy conversion system. This chapter describes what has to be done if this decision is to go ahead. Usually the installation will be under the direction of the supplier for machines greater than about 10 kW capacity. Likewise initial operation will be monitored by the supplier, who may indeed continue with maintenance provision for later years. Nevertheless the owner should be aware of what has to be done for installation and operation, and occasionally owners will be prepared to undertake much of the work themselves.

7.2 INITIAL CONSIDERATIONS

All stages of installation and operation must be fully confirmed before making a start. This may seem unnecessary, and is not the approach for normal purchases such as a car or tractor. However for a small wind energy conversion system (SWECS) the many components must work together, and a difficulty in one aspect, say the electrical connection, can spoil the whole system.

7.2.1 Planning permission

Planning permission must be obtained before installation, and usually detailed machine and site descriptions will have been given. Planning authorities may require periodic inspection of the project as it proceeds. Since the installation of SWECS is not common, it is important to have good relationships with the authorities since they will be learning also.

7.2.2 Electricity Boards

Electricity grid connected installations must have the permission of the local Board before connection is made, and the financial arrangements will also need to be finalised. According to the 1983 Energy Act, Boards can only, quite reasonably, refuse grid connection because of safety aspects or insuperable technical problems. If the machine is one of the first in the area, the Board may want to monitor performance for their own benefit. To do this they should be helpful to the owner, and may provide instrumentation. However all instruments and meters regarding the exchange of power must be paid for by the owner.

7.2.3 Machine supplier

Purchasers must take every precaution to check that the machine to be purchased is reliable and meets standards of construction and operation. A detailed specification should be provided by the supplier and this should be checked against national standards wherever possible. Be particularly careful about changes in original specifications that may not have been fully tested and approved. It must be emphasised that a wind turbine can be a dangerous device if it is not properly designed and protected. EEC and UK national standards of certification for wind turbines will be a manufacturing requirement for commercial machines, and customers should enquire about the latest standards.

7.2.4 Meteorological information

Determine from the local meteorological office the best available wind data for the site. Usually only general data are available, but it is important to know average wind speed and maximum gust levels. Check that the intended machine is suitable for these conditions.

7.2.5 Insurance

Insurance cover is very advisable. This may be obtained from your usual company, but many also require discussion with a specialist company. See Chapter 5 for more details.

7.2.6 Siting

Finally check that the proposed site is as you intended, and that there is access for delivery of all the components.

7.3 SITE PREPARATION AND INSTALLATION

In practice site preparation and installation of the machine tends to happen closely together. The work may be undertaken by the supplier, a separate contractor (but take care to find one who is experienced) or by the owner's own facilities. The latter will almost certainly be the cheapest so long as a proper standard is reached.

7.3.1 Site selection

Site selection criteria have been explained in Chapter 2. It will have been advisable to monitor the wind on the site for several months beforehand for comparison with the general meteorological data. If indeed an anemometer has been used, it is wise to continue taking measurements of wind speed after installation to check machine performance.

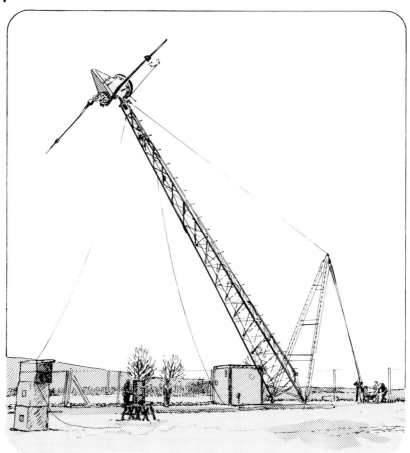

Fig. 7.1   Erecting a 15 kW, 6m diameter wind turbine

7.3.2 Erection

There are several methods for erecting wind turbines. These include:

** Manpower using shear legs or an A frame, and muscles. Several people are needed and this is more difficult than it seems, Figure 7.1.

** Handpowered winch and shear legs or A frame. This requires less manpower and is practical for quite large machines. It is also quick. See Figure 7.1.

** Direct lift using a crane. This is expensive unless a large crane is readily available.

** Tower erection followed by use of the tower as a crane to lift the head components. This requires few men but is slow.

Each of these requires a different approach, and may require additional foundations, e.g. for the winching post. The ground state may preclude heavy equipment being used for the operation, but this can be remedied using steel mesh mats.

7.3.3 Safety

Safety equipment, which may be hired, should be available during each phase of the work. This includes hard hats, and during the erection phase, safety-belts when working on the tower. Leather gloves will prevent hand injuries from the sharp remains of welding or galvanising on the tower structure. If welding is necessary on the job protective gear must be made available. Perhaps the greatest danger is from items dropped from the tower.

7.3.4 Site work

Preliminary preparations may entail the removal of trees which otherwise would reduce the wind speed at the rotor, and the treatment of hedges for the same reason. However, there may be protection orders on some trees and hedges. It is unlikely that buildings will be demolished to make way for the machine, but if this is necessary there is a ready supply of hard core.

7.3.5 Foundations

The size of foundation will be advised by the supplier but ensure that there are no special needs for the area, which will be advised by the local planners. Reference to local builders may give an indication of any special problem and they may be willing to prepare the foundation.

In the event that the purchaser is installing the machine, foundation preparation details may be found in Kemp's Engineer's Handbook and the Cement and Concrete Association's 'Man on the Job' leaflets (see Bibliography). Concrete suppliers may advise on the particular mix needed. Steel reinforcement will be needed as described by the turbine manufacturers.

There may be an advantage in using a depressed concrete foundation (i.e. lower than ground level), so allowing the area under the turbine to be regrassed (it seems cattle are unaffected by the movement and noise of the blades).

7.3.6 Electrical connections
Provision must be made for the conduits and cables carrying power and signals to and from the turbine. It may be necessary to lay these before pouring the concrete. The choice between overhead and buried cable will be dictated by the use of the site surrounds and the proximity of the installation to buildings, roads and possible earthworks (ploughing and trenching). If the site is readily accessible, warning signs should be prepared to highlight electrical and general dangers.

7.3.7 Restriction of access
Personnel and animals must be kept from dangerous aspects of the installation, although access to the base area may be permitted. Unauthorised climbing of the tower must be prevented and so the tower area should be fitted with anti-climb features similar to those used on electricity pylons. A full fence round the site may be required to prevent climbing or interference with cables. Such a fence also protects the public against items dropped during maintenance.

7.3.8 Co-ordination
It will be important to co-ordinate the various stages of the project, even if the buyer is proposing to undertake the installation. There may be several contractors involved for the excavations, foundations and installation. There may also be various authorities who have to approve features of the installation. It may be important to plan to install associated equipment, such as electrical loads, while the concrete foundations are curing.

Before commissioning the machine the site should be cleared of debris and unused equipment, and restored to the original state. Access should remain for possible emergencies.

7.4 COMMISSIONING

Success at this stage is crucial to the whole operation, otherwise there will be frustration, ill feeling and disillusionment. Establish early on who will be responsible for commissioning. For a small machine, commissioning will be straightforward, especially if there is no requirement for grid connection. A larger grid connected machine will need careful attention and the setting of various relays etc. Local grid conditions will be important and should not be assumed to be standard. The Electricity Area Board must be consulted.

A check-list (similar to a car dealer's check-list) should be available from the manufacturer. This is used to confirm that all the features have been aligned mechanically and electrically, and that the assembly functions correctly. It may be helpful to visit some other installations, preferably during erection. A photographic record would be useful. Operators are advised to have some training with the manufacturer as part of the agreement to purchase the machine.

A record system should be provided with the equipment (similar to a car maintenance book) so that the various services could be recorded. The requirements of each service should be provided, but these may not be as extensive as for a car. In some cases the manufacturer or supplier will not permit the buyer to service the installation in order to preserve warranty conditions. This may introduce additional costs to the project and should be noted at an early stage.

The initial commissioning time should be less than a day. Simulation of all the control functions should take place during the initial commissioning phase. A test of each safety feature must be simulated and there must be a means to repeat this at regular intervals. Other features of the installation should be checked at the same time so that faults caused by loads can be rectified. It will be desirable for the operator to visit in a selection of weather conditions ensuring that operation is safe from both mechanical and electrical aspects. A return visit immediately after a storm should take place to ensure that parts have not become loosened. This may only be necessary at the first occasion, since machines should have satisfactory fail safe protective mechanisms.

In the early stages of wind turbine technology this type of approach ought to form part of a regular round of visits. There will be differences between large and small installations in the type of attention required. Contact telephone numbers and addresses should be provided and a 24 hour coverage would be beneficial.

7.5 TURBINE MAINTENANCE

A general check-list should have been made available to the operator by the supplier and this will form the basis of the work. It is important that the checks are made regularly, preferably with the intervals being shorter rather than longer. It will be found that installations at high wind sites will be more difficult to maintain because of access. The times of low wind should be taken as periods of inspection and maintenance.

Attention should be paid to the structure, especially before and after the windy seasons. Weld and bolt conditions should be noted and also any effects of corrosion on the structure, particularly in a marine environment.

Few sites will require active anti-icing provisions on the rotor blades, although some machines may have treatment to the leading edge. In certain conditions the build-up of rime and ice can be considerable. This could result in the rotor being out of balance which should trip any vibration sensor, if present. A more serious condition would be a general build-up of ice on blades in an equal amount, not causing an imbalance, which could over-stress the rotating structure. Hopefully the profile of the blades would be spoiled sufficiently to cause a reduction in rotor speed. In these conditions, frequent inspection, coupled to the machine being shut down, would be the wise course to take. If during a period of shut down ice has been allowed to build up, this must be removed or allowed to thaw before the machine is started. This would apply to any size of machine, but the problem of handling a small machine may be more difficult.

During cold conditions other difficulties may be encountered, particularly in a compressed air circuit or lubrication system. Supplementary heat, moisture dispersants, and good housekeeping should overcome such difficulties.

For some major tasks the tower may need lowering, enabling safe access to the rotating components. Features for this should have been left behind from the initial erection procedure. Safety equipment should be kept as part of the maintenance kit and the temptation to do without must be resisted. An addition to the equipment should be warm clothing and mittens - wind chill factors can be considerable. Some of the larger machines may provide the means to lower components from the nacelle, this again will need assistance.

A list of the tooling necessary should have been provided and the tools themselves should be purchased before being needed. It is unlikely that there will be a requirement for special tools. At certain sites there may be some damage occurring due to vandalism, especially as a new piece of equipment may provide a temptation. Preventative measures should be maintained and the installation checked for stone and air rifle damage. Binoculars may be useful.

The site will have been chosen to be free from obstructions. However it may become congested later, due to tree and bush growth, or buildings may have been introduced which disturb the wind flow. Comparison with original photographs will show any changes that have taken place. Remedial action should be taken to restore or improve the original conditions and the resulting energy conversion.

7.6 LOAD OR STORAGE MANAGEMENT

A check of the electrical distribution system, including transmission poles, insulators and plugs must form part of maintenance. Electrical wiring should have been installed so that it is easily accessible for possible replacement. Rodents however may develop a taste for insulation and accessible areas should be protected.

The simplest load is likely to be heating, in the form of convective or radiant elements, or immersion water heaters. A check can be made (with the power off and suitable disconnections made) using a cheap multimeter to ensure the correct resistance is present. Provided the heating units have not been placed in concrete, it will be a simple matter to make a replacement. This should not be delayed since reduced resistance in the circuit will cause a general rise in current and a speedier failure in the remaining elements.

A battery is a convenient store for wind generated electricity and may form part of a system for any size of turbine. Batteries are frequently neglected, yet they form an expensive and important part of a system. Electrolyte levels will need regular checking. If glass sided cells are used this is simple, but removal of plugs from the opaque cased batteries is time consuming. Terminals should be greased to prevent corrosion. Further details are given in Section 4.5.

7.7 MONITORING THE SWECS AND ASSOCIATED INSTALLATIONS

The only way to monitor performance is by using instruments, so these must be incorporated in the system. The requirements will be dependent on the user, the type of installation and the loads which are being serviced. Some instruments are however expensive and should not be introduced without thought.

Instruments which permanently record data will be preferable to those which give an instantaneous reading. If some form of instant reading can be provided at little extra cost, this does add interest and provides a useful check on the operation of the data logger.

During the initial years of the project the supplier should have an anemometer to establish the site wind speed characteristics. Wind speed may vary significantly from one year to another and a period of low wind speeds may be the cause of lower than average output.

An installation of 5 kW capacity or more will benefit from detailed instruments, which would include monitoring of:

**   Power (kW) in a recorded and in an instantaneous form

**   Reactive power (kWAr) as above

**   Turbine running hours

**   Turbine generating hours

**   Current measurement (not so important if the power is being measured, particularly in a fixed voltage situation)

Active and reactive power records are important in order to check the local Board's charges in the case of a grid connected machine. The various running times will indicate the amount of service activity required and the average output of the installation. A full set of instrumentation will provide a basis for estimating the performance of the turbine.

If a complex electrical demand is being supplied, it may be necessary to examine how the turbine interacts with the load. Possibly the supply does not match the demand, and a better stratagem needs to be used for operating the wind generator. This sort of matching should have formed part of the initial study for the installation, but the demand on the system may have changed and a revision could improve performance.

7.8   OVERHAUL AND PROVISION OF SPARES

The tasks undertaken during the commissioning and maintenance phases will have given good grounding for the general overhaul of the turbine and the installation. There should be several years of operation before wear becomes a problem. The need for an overhaul will depend upon the usage of the turbine, the number of stops and starts, the hours of operation and the rigours of the environment. It may be preferable for the supplier to undertake these tasks, especially if the machine needs to be lowered or lifting tackle is required. If so this could have formed part of the service contract.

Ensure that warranties are not invalidated in tackling the job oneself, or in getting an unauthorised installer to do the work. Spares must be available before major tasks are begun. Machines may have been imported and parts may be restricted in supply. Major components will need a significant delivery time and these should be budgeted for in advance.

7.9   TROUBLE-SHOOTING

A variety of obvious faults (to the other person!) will cause a failure and some general items are listed here.   Fault diagnosis should be listed in the turbine manufacturer's manual.   Some faults may be specific to the particular model.   Do not be tempted to overlook the obvious and in the event of complete frustration discuss the problem with someone else to bring out the missing points.

General faults include:

**   The circuit.
        NB if any parts are live, disconnect the power.
        Is it continuous?
        Look for breaks (excavations, weather damage, etc.)
        Are the fuses intact?
        Have some relays tripped?

**   Instrumentation.
        If relays are included what is the state of the indicator lights?
        Are there any indications of current or voltage?
        Do resets need restoring after a temporary fault?

The following are guide-lines for non-operation of the machine:

        Check the status of the display lights.
        Is there or has there been a power cut?
        Fuses have failed - why?
        Cable failure?
        Load fault?
        Resets have not been made?
        Vibration monitor tripped? Reset and run machine up
         carefully to see why.
        Ice buildup on the blades?
        Bird strike?
        Relays have been tampered with?

7.10 QUICK REFERENCE GUIDE : A SUMMARY
7.10.1 Initial considerations
Before commencing installation, the following organisations may need to be contacted:

a   Local Authority, to confirm planning permission.

b   Electricity Board.

c   Supplier, to confirm the specification of the machine being supplied, and that any modifications are according to the turbine authorising authority and the Electricity Board.

d   Installation contractor, to ensure that the timings for equipment and manpower are as arranged, or that amendments can be made. Take care that the foundations are adequate for the specific location and ground conditions.

e   Local meteorological office, to confirm the levels of extreme winds and verify that the supplier is confident that the installation is able to withstand the conditions.

f   Insuring agents, to confirm the installation is insured and conforms to their requirements.

A review of the following points would be prudent:

a   Site inspection to confirm the wind fetches have not been obstructed and that all potential hazards have been dealt with (flying club, low level flight zone, bird movements, pollution source).

b   Access routes to site are suitable for the transportation of the equipment and that road or bridge construction does not constrict access.

c   People who may find the turbine an annoyance (noise, visual intrusion etc.). Are the original permissions still valid?

d   Is there a wind turbine users' association to join?

e   Electrical properties of the installation - have there been any major changes on the local feeder which the installation may affect or which may affect the installation?

7.10.2 Siting
Refer also to Chapter 2.

a   Prepare schedules and confirm with the supplier, to ensure that manpower and equipment are not hired unproductively.

b   Prepare the site area.   If the area is soft, ensure that temporary surfaces are installed, and clear trees if interfering with expected wind flows.   (Are the trees 'protected'?)

c   Prepare the foundations.   Allow ample time for the foundations to set.

d   Prepare conduits for cable runs (power, controls and instrumentation).

e   Confirm the methods to be used for the erection of the machine (crane, winch etc.).

f   Ensure that safety precautions are adequate during the erection phase - hard hats, harnesses, gloves etc.

g   Electrical safety - warnings - deterrents.

h   Protective measures around the installation, restricted access.

i   Site clearance following the installation, but leaving provision for access by emergency services.

j   Ensure that the associated installations of equipment are keeping to the schedule.

7.10.3 Commissioning

a   Provision of check-lists by the manufacturer to the supplier or to the new owner.

b   Who will undertake the commissioning - manufacturer, supplier or self?

c   Is there a training scheme required?

d   Manufacturer's guarantee: a check for validity if some of the maintenance is undertaken personally.

e   Maintenance of a record system (car log-book type?) for use in customer/user disputes.   Keep careful written records.

f   Testing of control functions, particularly those which are less likely to be used, e.g. the overspeed control.

g   Simulation of failure conditions so that safety procedures may be validated.

h   Ensure some regular visits from the manufacturer's agent during a variety of conditions so that the shake down may be comprehensive.

7.10.4 Maintenance

a   General check-lists: it is important that the various checks are made at the correct frequency.

b   Structural inspection: to include weld/bolt integrity, corrosion (particularly in a marine environment).

c   Inspections following severe weather.

d   During low temperature conditions note the condition of the rotor blade with respect to possible ice and hoar frost accretion, and if present allow this to dissipate before operating the machine.

e   Does the tower require lowering for any of the work? Can the task be undertaken with few personnel easily?

f   Availability of tooling, lubricants, and skilled labour if necessary.

g   On hand safety equipment (hats, harnesses, instruction).

h   Note and remedy any effects of vandalism.

i   Ensure that the site is kept free of obstructions to the wind, particularly after some years of operation - tree growth, building development (a photographic record of the initial site is valuable). Changes in land use nearby in respect to the type of crops grown may reduce the rotor height wind speed.

7.10.5 Monitoring and instrumentation
Requirements will vary dependent upon the needs of the user and size of the installation.

a   A multimeter is the most basic test instrument for voltage and line continuity.

b   Full instrumentation packages could include monitoring the power (active and reactive), current, frequency, running hours and usage of the output.

c   Possible instruments might include:
  Light display, indicating generation.
  Anemometer, plus a display for the average wind speed.
  Power and reactive power meters.
  Monitors of energy usage (integrating power meters).
  Ammeter.
  Installation state by light indicators.

d   A full set of instruments will monitor the machine performance and be invaluable to check both operation and faults.

7.10.6 Overhaul and provision of spares

a   Is the overhaul to be completed by the manufacturer, supplier or the owner?

b   Is a service contract available?

c   Are instructions given for do-it-yourself servicing? Would this invalidate warranties and insurances?

d   Spares will need to be imported for foreign machines so the crucial items will need to be held by the supplier. Confirm that other items can be supplied within a reasonable time at a reasonable cost. Will the supplier provide a machine to cover the waiting period?

e   Confirm the availability of spares for locally manufactured machines. (Some of these items may be imported.)

7.10.7 Trouble-shooting

The following guide-lines will be useful for non-operation of the machine:

Check the status of the relay lights.

Is there or has there been a power cut?

Fuses have failed - why?

Cable failure.

Load fault.

Resets have not been made.

The vibration monitor has tripped, reset and run machine up to see why. Ice buildup on blades? Bird strike?

Relays have been tampered with.

7.10.8 Maintenance of loads or storage medium

The main loads are likely to be checked as part of a general site maintenance programme, but if a battery store is part of the system this will need regular attention, as will other storage media. It is essential to keep batteries in good condition and away from possibilities of short circuits.

7.10.9 Economics of ongoing operation

Using the installed meters it will be possible to estimate the levels of performance of the installation and confirm the original estimates. Chapter 6 covers the economics of operation and some remedial action may be indicated.

8   WIND POWER FOR PUMPING WATER

Contents

8.1   INTRODUCTION

Apart from milling, the traditional use of wind turbines has been for pumping water. Throughout the world people will have seen wind turbine water pumps, usually for cattle troughs or for irrigation. Although this Guide Book mainly concerns electricity generating machines, it is important that water pumping turbines are also discussed, albeit briefly.

The main type of wind-pump that has been used is the so-called American farm wind-pump, see Figure 8.1 This normally involves a many-bladed rotor having steel cambered plates arranged like a fan, which drives a reciprocating linkage usually via reduction gearing that connects with a piston pump located in a borehole directly below. This design evolved between 1860 and 1900 when many millions of cattle were being introduced on the North American prairies. Limited surface water created a vast demand for water lifting and wind-pumps were rapidly identified as the best general power source for this purpose. The farm wind pump reached its zenith in terms of numbers in use in the U.S.A. during the early 1920s, by which time some six million had been installed. Rural electrifiction in the U.S.A. led to a decline in the use of the wind pumps, to an estimated 150,000 today.

Other new frontiers such as Australia, Argentina and South Africa took up the farm wind-pump, and to this day an estimated one million steel farm wind-pumps are in regular use, the largest numbers being in Australia and Argentina. It should be noted that the so-called American farm wind-pump is only rarely used for irrigation; the majority are used for the purpose they were originally developed for, namely livestock and, to a lesser extent, human water supplies. They tend therefore to be applied at quite high heads by irrigation standards; typically in the 10 - 100 m range on boreholes. Large wind-pumps are in use on boreholes of over 200 m depth.

Some 50,000 wind-pumps were used around the Mediterranean Sea some 40 years ago for irrigation purposes. These were commonly improvised direct-drive variations of the metal American farm wind-pump, but often using cloth sails rather than metal blades. These cloth sail windmills are often called Cretan windmills. This type of rotor has been used for several centuries in the Mediterranean region for corn milling windmills. During the last 30 years or so, increased prosperity combined with cheaper engines and fuels has generally led farmers in this region to abandon wind-pumps and use small engines (or mains

lectricity where available).   However, Crete is well known as a
ountry with about 6,000 wind-pumps still in use, mostly with the cloth
ailed rotor.

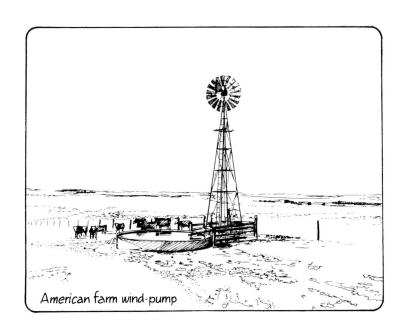

American farm wind-pump

Fig. 8.1   American farm wind pump.

8.2 BASICS OF WATER PUMPING

There are two very distinct uses for wind-pumps,
i) providing drinking water for people and animals, (ii) irrigation.
These give rise to two distinct categories of wind-pump because the
echnical, operational and economic requirements are generally
different.

Drinking supply wind-pumps need to be reliable with the minimum of
maintenance, to run unattended often remote from human habitation and to
pump water generally from depths of 10 m or more. A typical farm wind-
pump will run for at least 20 years or more without any major components
needing replacement and with maintenance only once every year.   This is
a very demanding technical requirement since a typical wind-pump should
operate for about 75,000 hours before anything significant wears out.
This is 4 - 10 times the operating life of diesel engines or over 20
times the life of many small engine driven pumps. Wind-pumps to this
standard therefore are industrially manufactured from steel components
and drive piston pumps via reciprocating pump rods.   Inevitably they
are quite expensive in relation to their power output because of the
heavy-duty nature of their construction.   However a high premium is
worth paying to achieve reliability and minimum need for human
intervention, as this is the wind pump's main advantage over practically
any other form of pumping system.

On the other hand wind turbines for irrigation usually have only seasonal use. These pump large volumes of water through a low head and the intrinsic value of the water is low. Therefore obtaining low costs must override most other considerations. Since irrigation generally involves the farmer and/or other workers being present, it is not so important to have a machine capable of running unattended. Therefore windmills used for irrigation in the past tend to be indigenous designs that are often improvised or built by the farmer, such as in Crete or China, as a method of low cost mechanisation. Standard farm wind-pumps are usually unsuitable for irrigation because low head, high volume pumping is not suited to piston pumps. Also, most wind turbines for drinking water supply have to be located directly over the pump or substantial reinforced concrete foundations which usually makes such machines unsuitable for pumping from open water. Therefore most irrigation wind-pumps use rotary pumps which are more suitable for low heads. Usually such systems do not require concrete foundations, and the tower is kept low. Consequently they are cheaper and easier to install than drinking water supply systems.

Attempts have been made recently to develop low cost steel wind-pumps. Traditional wind-pump designs, even though still in commercial production, generally date back to the 1920s and are therefore heavy and expensive to manufacture by modern standards and difficult to install properly in remote areas. However small-scale manufacturing techniques today are quite different. Welded steel fabrication is cheaper than bolting together heavy forgings and castings, grease lubricated rolling bearings are mass produced and allow simple mechanical design, and much is understood about both rotor aerodynamics and system design. These improvements allow direct drive wind-pumps to be used where previously reduction gearing was necessary. Therefore it has become possible to revise the traditional farm wind-pump concept into a lighter and simpler modern form (see Figure 8.2). Costs may now be kept low enough to allow the development of all-steel durable wind-pumps, that are cheap enough for irrigation.

8.3 MATCHING ROTORS TO PUMPS

High solidity rotors, which are rotors usually consisting of numerous blades set like a fan, are typically used in conjunction with positive displacement (piston) pumps. This is because a many-bladed, high solidity rotor has a starting torque two to three times the running torque. This happens to suit piston pumps since they need about three times as much torque to start as to keep going. Low solidity (propeller type) rotors are best for use with centrifugal pumps, ladder pumps, and chain and washer pumps, where the torque needed by the pump for starting is less than that needed for running at design speed.

The difficulty in designing wind turbine pumps is that the volume pumped depends directly on the pumping frequency, which itself is proportional to the wind speed. However the power developed by the turbine rotor increases as the cube of the wind speed and hence the cube of the rotational frequency for an efficient rotor. It is not therefore possible to maintain good efficiency for a range of wind speeds using the usual pumps directly coupled to the rotor.   This mismatch is actually less serious than it may seem, since the time when the best efficiency is needed is at low wind speeds.   When the rotor is running fast enough to be badly matched with its pump, the output is reduced by bad matching but still usually more than adequate.

Centrifugal pumps can in theory be made to match well with a wind turbine, but their efficiency falls rapidly to zero below a certain threshold running speed at a given static head.   Therefore matching problems have actually proved more of a problem with centrifugal pumps directly driven by wind turbines than with positive displacement pumps. The latter, although poorly matched at high speeds, do produce a good output at low speeds where efficiency is important.

There is considerable scope for improving the overall performance of wind-pumps by developing methods of improving the rotor-to-pump match over a wider range of wind speeds.   Research is being carried out and, if successful, could result in considerably more effective wind-pumps in the future

Fig. 8.2 Modern wind turbine pump

The operating characteristic of a typical wind-pump is shown in Figure 8.3. This shows that it starts in wind speed v_s (the start up wind speed) but can run down to a slightly lower wind speed v_{min} when coupled to a pump with a lower running torque than its starting torque. The pump reaches its best match with the rotor at wind speeds close to v_{min} and then increases its output almost linearly with wind speed to v_r (its rated wind speed). At still higher wind speeds, devices prevent the rotor speeding up further to avoid damage. At very high wind speeds, the only safe course of action is to furl the windmill completely. Figure 8.3 shows how this process commences at a wind speed v_f (furling speed) and is completed at wind speed v_{sd} (shut down).

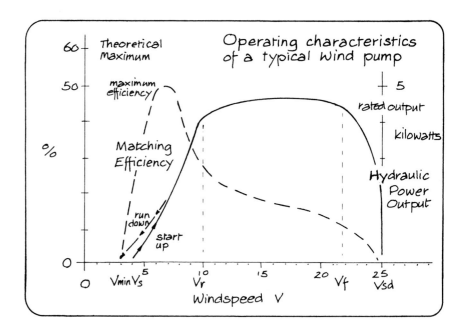

Fig. 8.3 Operating characteristics of a typical wind-pump

8.4   METHODS OF STORM PROTECTION AND FURLING

Windmills must have a means to limit the power they can deliver, or else they would have to be built excessively strongly (and so be expensive) merely to withstand very occasional high power outputs. The traditional method of achieving this with sail windmills and other such simple traditional designs is to reduce the sail area. Metal farm windmills, however, have fixed steel blades, so they generally use an arrangement in which strong winds cause the rotor to yaw edge-on to the wind.   This is arranged by mounting the rotor offset from the tower centre (see Figure 8.4) so that the wind constantly seeks to turn the rotor behind the tower. Under normal conditions the rotor is held into the wind by a long tail with vane on it.   This vane is hinged and restrained, either with a pre-loaded spring (as illustrated) or by having its hinge inclined.   When the wind load on the rotor increases the tail will start to fold and the wind pushes the rotor around to present its edge to the wind.   This furling process starts when the rated output is reached.   If the wind speed continues to rise, the movement increases progressively until the machine is fully furled and the rotor ceases to rotate. When the wind drops, the tail unfolds and the wind on the tail turns the rotor once again to face the wind.   This action is completely automatic, but may also be operated manually during times of maintenance.

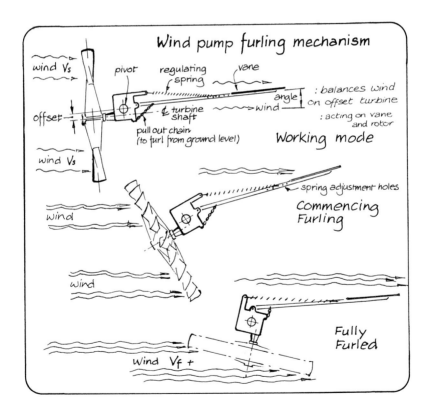

Fig.8.4   Furling mechanisms

8.5 WIND-PUMP MANUFACTURERS' PERFORMANCE CLAIMS

The easiest method of estimating the performance to be expected from a wind-pump is to use manufacturers' data as printed in their brochures. For example, Figure 8.5 shows performance curves for the Kenyan made 'Kijito' range of wind-pumps. The table indicates the average daily output to be expected at different pumping heads for the four sizes of Kijitos in three different average wind speeds, defined as 'light', 2-3 m/s, 'medium' 3-4 m/s, and 'strong' 4-5 m/s. The curves reproduce these results just for the 'medium' wind speeds. It is interesting to note how sensitive wind-pumps are to wind speed; the smallest machine with a 3.7 m rotor will perform in a 5 m/s wind almost as well as the largest machine (7.3 m) does in a 3 m/s wind. This is because there is 4.6 times as much energy per unit cross-section of a 5 m/s wind as in a 3 m/s wind as a result of the cube law, equation 3.1.

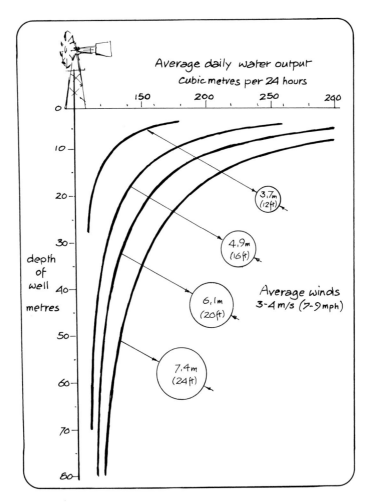

Fig. 8.5 Average daily water output (cubic metres per 24hrs) for winds averaging 3-4 m/s (7-9mph) plotted against depth of well (metres).

A simple rule of thumb to estimate the energy output from a windpump is to use the formula $E = 0.1\ v^3At$, where v is the mean wind speed, A is the area of the wind rotor and E is the hydraulic energy output for the period t. For example an average wind speed of 3 m/s, with a wind turbine rotor of 3.6 m diameter ($A = 10m^2$), will yield approximately in 24 hours

$$E = 0.1 \times 4^3 \times 10 \times 24 = 1536 \text{ Wh}$$

A problem with manufacturers' performance claims is that some brochures include inaccurate, unreliable or even incomplete data.   For example, it sometimes is not clear what wind speed applies for the manufacturers' claimed outputs.   There is a tendency for manufacturers to quote performance figures for unusually high average wind speeds, no doubt because this makes the performance look more impressive, and they then give rules of thumb which in some cases do not seem accurate for reducing the outputs to more realistic levels for more common wind speeds.

The difficulty inherent in monitoring the performance of a wind-pump often prevents users from actually checking whether they are getting what was promised. If a cup counter anemometer and a water meter is available, it is possible to measure the wind run and the water output over fixed periods of time (either 10 minute intervals or daily intervals both short-term and long-term is recommended).   The mean wind speed and mean water output over these periods can be logged and then plotted on a graph.   When enough points are obtained on the scattergram, a 'best fit' curve can be drawn to obtain the performance characteristic.   The results should then be compared with the manufacturer's claims.   In some cases sub-optimal performance can occur simply because the wrong pump size or wrong stroke has been used, and considerable improvements in performance may result from exchanging the pump for the correct size or from altering the stroke.

9   CASE STUDIES

Contents

9.1   INTRODUCTION

Since this Guide Book is intended to encourage the application of wind energy in the United Kingdom, a chapter describing a selection of operational installations is important.   Research machines and test sites have been excluded as the intention is to focus on the application of commercial wind turbines to power generation, or, say, water pumping.   Machines of capacity greater than 100 kW have been excluded as beyond the remit of this book.

The examples chosen can all be considered successful but we do not disguise the fact that in some early applications there have been a number of machine failures.   The BWEA is not suggesting that the wind turbines described are necessarily the best available or to be especially recommended.   They have been chosen primarily because information on actual operation was available.   It is hoped that the cases covered will give prospective purchasers both ideas for applications and increased confidence in the ability of wind energy to satisfy certain forms of energy requirement.

Wind energy deployment in the U.K. started slowly as compared with parts of Europe and the U.S.A., where the scale of electricity generation on the wind farms of California has been impressive.   Some explanation for the slow initial progress in the U.K. was provided in Chapter 1. However wind energy can now be considered competitive with many conventional energy supplies for a range of applications in the U.K. Early deployment occurred where the cost of conventional energy is high, as with electricity on islands not supplied by the nationl grid.   The usual means of generation in these circumstances is by generators powered by diesel engines. Depending on the local conditions, electricity costs can be as high as 20p per kWh.   A similarly clear cost advantage for wind power can exist for specialist applications such as battery charging.

Encouraged by the 1983 Energy Act, private generation into the grid has become acceptable, although it must be stressed that the economics here are heavily dependent on the tariff structure set by the relevant area Electricity Board (see Chapter 5). In these situations due regard should also be paid to reactive power consumption for which the local Board will make charges. (Reactive power occurs when voltage and current are not in phase, and is usually due to motors or generators dominating the local network). On balance it appears that with the present tariff arrangements it is in the interests of the wind turbine operator to maximise the local consumption of wind generated electricity rather than sell to the Electricity Board concerned. In these cases careful management of the user's load to coincide with generation from the wind will certainly be worthwhile.

Essential details of the case studies are listed in Table 9.1. Figure 9.1 shows their geographical location. Where possible the presentation of the case studies has been fitted into a common format to facilitate comparison.

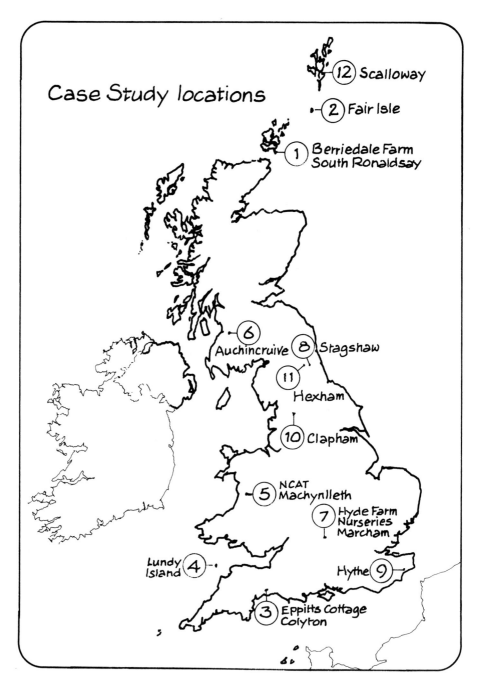

Fig.9.1 Location of the case studies

CASE STUDY	NAME	TURBINE CAPACITY (watts or kilowatts)	PURPOSE AND COMMENT
1	Berriedale Farm, South Ronaldsay, Orkney	22 kW	Grid connected. Experience for Electricity Board.
2	Fair Isle Shetland	55 kW	Remote community mains electricity.
3	Eppitts Cottage, Colyton, Devon	500 W	Wind/battery lighting system.
4	Lundy Island, Bristol Channel	55 kW	Remote community mains electricity.
5	National Centre for Alternative Technology, Machynlleth, Powys	15 kW	Household heating and 240V electricity lights. Battery back-up.
6	West of Scotland Agricultural College, Auchincruive, Ayr	15 kW	Grid connected. Experience for Electricity Board.
7	Hyde Farm Nurseries, Marcham, Oxfordshire	200 W (mech.)	Mechanical water pump.
8	Stagshaw Northumberland	1 kW	Wind/diesel system for household heating.
9	Hythe Folkestone, Kent	6 kW	Heating and electricity supplies for a house.
10	Clapham, Lancashire	500 W	Low voltage household electricity.
11	Brunton Bank Hexham, Northumberland	100 W	Low voltage household electricity and petrol generator.
12	Scalloway Shetland	55 kW	Private grid connected, for mains electricity use and sale to grid.

Table 9.1   Details of Case Studies

9.2.1   CASE STUDY 1 Berriedale Farm, South Ronaldsay, Orkney

The North of Scotland Hydro-Electric Board installed and commissioned a 22 kW wind turbine generator at Berriedale Farm on South Ronaldsay in December 1980.   Operation has continued successfully, without major problems since that date, although some attention was needed to the fan tails. In February 1984, 7 kVAr of reactive power factor correction was installed which reduced significantly the reactive kVAr absorbed by the machine.   By the end of 1984 a total generated output of approximately 200 MWh was reported.   The installation has been monitored by the Hydro Board to improve the understanding of such grid connected machines.   This experience will prove valuable with the expected growth in wind energy generation on the Scottish Islands. Maintenance and supervisory requirements have been extremely low, and the Board is well satisfied with the system.

EARLY REPORT   Pope,I T, "North of Scotland Hydro-Electric Board experience with 22 kW aerogenerator". Energy for Rural and Island Communities II, ed. J W Twidell, Pergamon, Oxford, 1982.

Holding, N L,"Small aerogenerators on farms", Proc. of BWEA meeting on "The Energy Act", University of Strathclyde, Glasgow, 1984.

SITE   Aspect: open and flat
               Mean wind speed: 8 m/s

TURBINE   Rotor: 3 bladed, fixed pitch
               10.2 m diameter
               Upwind
               Yawing by twin fan tails

               Rated output: 22 kW
               Rated wind speed: 12 m/s
               Cut in wind speed: 4.5 m/s
               Maximum shaft speed: 69 rpm

TOWER   Galvanised steel lattice
               Tower height: 18 m
               Hub height: 18.5 m

ALTERNATOR   3 phase, 415 V, 50 Hz, induction generator

SUPPLIER   International Research and Development Co Ltd.
               Fossway, Newcastle upon Tyne, NEG 2YD

OPERATOR   North of Scotland Hydro Electric Board

Fig.9.2 Berriedale Farm, South Ronaldsay, Orkney

OPERATING PERFORMANCE SUMMARY	1980/81	1981/82	1982/83	1983/4
Wind turbine output (kWh)	44,280	66,400	63,840	24,600
Reactive power consumed (kVArh)	63,730	48,420	62,000	6,180
Annual load factor (%)	21	34	33	25
Average power output (kW)	4.5	7.5	7.5	5.5

9.2.2 CASE STUDY 2 Fair Isle, Shetland

Fair Isle is a very isolated island in the North Sea with a population of about 70. It is famous as a bird sanctuary and is owned by the National Trust for Scotland. Most islanders live to the south of the island where electricity is supplied to 20 houses by underground cable. By 1980 the diesel supplied electricity was becoming too expensive, and so a 55 kW wind turbine generator was sought by the Island Electricity Co-operative. Advice was obtained throughout the project from the North of Scotland Hydro-Electric Board, and the system was designed and installed by the International Research and Development Co Ltd of Newcastle. The Co-operative owns and operates the system under the management of a qualified engineer.

The whole system was innovative and has been of great significance because of the high operational efficiency and the large proportion of fossil fuel substituted. The control system consists of electrical loads, mostly water heaters and storage space heaters, that are switched on and off to regulate the turbine rotation. When the wind is strong, many loads are automatically switched on. When the wind speed is low, heating loads are automatically switched off and so only essential loads, such as low power lights, are on. If there is no wind, the diesel system is operated only for guaranteed hours in the evening and early morning.

After initial gearbox difficulties, the system has operated well and almost continuously. The greatest benefits have been the supply of electricity outside the guaranteed hours, the abundant supply of heat to houses previously cold and damp, and the low cost of the supply. Since 1984 the average cost of electricity has been about 3.2 pence/kWh (including diesel back up) and the cost of wind generated electricity about 2 pence/kWh. This is the cheapest publicly supplied electricity in the U.K. The Co-operative regularly adjusts the pricing tariff to ensure loads are enabled to present control for the machine and to minimise dumped power.

Fig. 9.3 Erecting the 55 kW wind turbine on Fair Isle,Shetland

EARLY REPORT   Sinclair B A, Stevenson W G and Somerville W M
                    "Wind power generation on Fair Isle", in Proc.Energy for
                    Rural and Island Communities III, ed J W Twidell et al,
                    Pergamon Press, Oxford, 1984.

SITE: Aspect: open, but with considerable cliffs
                    within 2 km. The site is therefore subject to
                    wind turbulence and extreme wind speeds.
                    Mean wind speed: 9 m/s
                    Maximum gusts expected: 55 m/s

TURBINE   Rotor: 3 bladed, fixed pitch
                    14.0 m diameter
                    Upwind
                    Yawing by fan tails

                    Rated output: 55kW
                    Rated wind speed: 13 m/s
                    Cut in wind speed: 5 m/s

TOWER   Galvanised steel lattice
                    Hub height   15 m

ALTERNATOR   3 phase, 415v, 50Hz, synchronous generator

SUPPLIER   International Research and Development Co.Ltd.
                    Fossway, Newcastle upon Tyne, NE6 2YD

OPERATOR   Fair Isle Electricity Co-operative
                    (Plant Engineer: Mr B Sinclair)

OPERATING PERFORMANCE	1981/82	1982/83	1983/84	1984/85	1985/86
Diesel:					
Hours run	2,970	2,001			
Generation (kWh)	49,872	37,295			
Wind turbine generator:					
Hours run	–	4,071			
Generation (kWh)	–	92,627			
Ratio of operational hours					
wind turbine/diesel	–	2.03			
Increase (compared to 1981/82) in					
Hours of generation	–	2.04			
Energy use	–	2.61			

9.2.3 CASE STUDY 3 Eppitts, Southleigh, Colyton, Devon

In 1981 Michael Murdy purchased a Tornado 500 W wind generator to provide electricity for his cottage in rural Devon. Although less than half a mile from the central electricity supply a connection cost of over £10,000 would have been involved. Wind energy appeared, overall, a more sensible solution. A diesel generator set which he had previously used was proving too expensive to run and maintain, and had the added disadvantage of being very noisy. At present the wind generator, backed up by 2 kWh of nickel-cadmium batteries, provides a 110 volt d.c. supply used for lighting. Should the batteries run down during prolonged periods of low wind, they can be charged via a battery charger from the diesel set. Since the introduction of the wind turbine the consumption of diesel fuel has dropped to about one fifth, saving at least £200 per annum.

SITE
: Aspect: towards top of a south facing hill
Mean wind speed: 4 m/s

TURBINE
: Rotor: 2 bladed high speed aerofoil
Epoxy impregnated wood construction
2.1 m diameter
Multi-vane starter propeller and spinner
Upwind
Yawing by tail vane

Rated output: 500 W
Rated wind speed: 11 m/s
Cut in wind speed: 3.6 m/s

TOWER
: Galvanised modular steel lattice
Tower height: 10 m

ALTERNATOR
: 3 phase, permanent magnet. Brushless, with silicon
diode recitification.
Directly coupled to rotor
100 V d.c. output

SUPPLIER
: Tornado Wind Generators Ltd.
75 Benslow Lane, Hitchin, Herts, SG4 9RA

OPERATOR
: Mr Michael Murdy,
Eppitts, Southleigh, Colyton, Devon, EX13 6JB

Fig. 9.4 Eppitts, Devon

9.2.4 CASE STUDY 4 Lundy Island, Bristol Channel
          Until November 1982 electricity on Lundy Island was
provided solely by small diesel generator sets. Since that date the
bulk has been supplied by the 55 kW wind turbine generator installed on
the island, resulting in the saving of significant amounts of diesel
fuel. The installation, which is owned and run by the Landmark Trust,
for the National Trust, utilises a load control system developed by the
International Research and Development Co.Ltd., who also installed the
wind turbine (see also Section 9.2.2, Fair Isle).

EARLY REPORT   Somerville, W M and Puddy, J Wind power on Lundy Island,
               Proc. of BWEA 5 workshop, ed. P J Musgrove, Cambridge
               University Press, 1983

               Infield, D G and Puddy, J, "Wind-powered electricity
               generation on Lundy Island", Energy for Rural and Island
               Communities III, ed. J Twidell et al, Pergamon Press,
               Oxford, 1984.

SITE   Aspect: open, flat island top
               Mean wind speed: 6 m/s

TURBINE   Rotor: 3 bladed, fixed pitch
               14.5 m diameter
               Upwind
               Yawing by twin fan tails

               Rated output: 55 kW
               Rated wind speed: 15 m/s
               Cut in wind speed: 4.5 m/s
               Maximum shaft speed: 50 rpm

TOWER   Galvanised steel lattice
               Tower height: 14.5 m
               Hub height: 15 m

ALTERNATOR   3 phase, 415 V, Markon synchronous generator

SUPPLIER/   International Research and Development Co.Ltd.,
SYSTEM   Fossway, Newcastle upon Tyne, NE6 2YD
DESIGNER

OPERATOR   Mr John Puddy (Engineer) for Landmark Trust,
               Lundy Island, Bristol Channel

Fig.9.5   IRD installation on Lundy Island

OPERATING PERFORMANCE   (Typical annual data)

Total hours of wind generation	5,500
Total wind turbine output (kWh)	145,000
Mean output of wind turbine (kW)	26
Total island consumption:	
Delivered electricity (kWh) includes diesel)	41,000
Wind turbine output (kWh)	40,000
Excess wind energy to the dump (kWh)	32,000

9.2.5   CASE STUDY 5 <u>National Centre for Alternative Technology,</u>
<u>Machynlleth, Powys</u>

A 15 kW Polenko, type WPS 10, was installed at Llwyngwern Quarry, the site of the National Centre for Alternative Technology, in the summer of 1984. It generates electricity mainly for heating, for some battery charging, and also supplies 50 Hz electricity for lighting and appliances.   A controller ensures that frequency is controlled by load switching. Annual output is 20,000 kWh.

EARLY REPORT   Todd R and Kirby T "A 15kW stand-alone windpower system".
In the 1986 Proc. of BWEA, pages 87 to 94, Mechanical
Engineering Publications, London, ed. M B Anderson and
S J R Powles.

SITE   Top of a steep hill,
mean wind speed at 10 m   5.6 m/s.

TURBINE   Rotor: 3 bladed, fixed pitch, 9.6 m diameter
Upwind, yawing by twin fan tails
Speed control by powered flap and load control
Rated output: 15 kW
Rated wind speed: 12 m/s
Cut in wind speed: 4 m/s
Cut out wind speed: 27 m/s
Gearbox ratio: 1:30.44

TOWER   Tubular steel
Hub height: 20 m

ALTERNATOR   3 phase, synchronous generator

MANUFACTURER   Polenko bv,
Remmerden 9, 3911 TZ Rhenen, Netherlands

OPERATOR   National Centre for Alternative Technology,
Llwyngwern Quarry, Machynlleth,
Powys, SY20 9AZ

Fig. 9.6 National Centre for Alternative Energy

OPERATING PERFORMANCE

Typical monthly output   1500 kWh
Best coefficient of performance   0.28

9.2.6 CASE STUDY 6   West of Scotland Agricultural College, Auchincruive, Ayr

In the autumn of 1984 a 15 kW wind turbine generator was installed in the grounds of the West of Scotland Agricultural College. It is owned and operated by the South of Scotland Electricity Board (SSEB) to gain experience in the operation of wind turbines and to allow data collection on performance and operating costs. The Board will assess the value of such machines and thereby be able to advise consumers who may wish to install similar machines. The machine was designed, manufactured and erected by International Research and Development Co Ltd, a wholly owned subsidiary of Northern Engineering Industries plc to the order of the SSEB. It is connected to the local electricity distribution network at the farm substation, and was commissioned in August 1984.

SITE	Open aspect field, top of a gentle ridge.
TURBINE	Rotor: 3 bladed, fixed pitch 7.9 m diameter Upwind Yawing by twin fan tails

Rated output:	15 kW
Rated wind speed:	11.5 m/s
Cut in wind speed:	4 - 5 m/s
Maximum operating wind speed:	30 m/s
Maximum power output:	19 kW at 20 m/s
Turbine rotor speed:	93.8 - 96.2 rpm
Generator speed:	1500 - 1533 rpm
Turbine overspeed trip:	110 rpm

TOWER	Tubular steel	
	Tower height:	12 m
	Hub height:	12.5 m
ALTERNATOR	3 phase, 415 V, 50 Hz, 4 pole, induction generator Frame D200L, T.E.F.V.	
MANUFACTURER	International Research and Development co Ltd. Fossway, Newcastle upon Tyne, NE6 2YD	
OPERATOR	South of Scotland Electricity Board	

Fig. 9.7   West of Scotland Agriculture College, Auchincruive, Ayr

9.2.7 CASE STUDY 7 Hyde Farm Nurseries, Marcham, Oxfordshire

A 'Climax' wind pump was installed at the nurseries near Marcham in about 1915. Since then at least 40 years of operation have been accumulated. In 1980 both the wind turbine and the pump received a major overhaul; new blades and a tail fin were fitted to the original specification. The wind pump now functions reliably but presently only pumps water through a 2 m head, from a natural reservoir to decorative pools. A 2,000 gallon water storage tower still exists as part of the nursery's water supply and it is hoped to use the wind pump to fill this as has been done in the past. This will mean a head of 12 m, which is well within the manufacturer's specification.

SITE	Aspect: generally open, flat	
	Mean wind speed:	3.5 m/s
TURBINE	Rotor: many-bladed	
	2.0 m diameter	
	Upwind	
	Yawing by tail vane	
TOWER	Galvanised steel lattice	
	Tower height to hub	10 m
PUMP	Single action oil bath type	
	Stroke:	178 mm (7 inches)
	Bore:	51 mm (2 inches)
	Maximum head:	38.1 m (125 feet)
MANUFACTURER	Wyatt Bros (Whitchurch) Ltd.	
	Wayland Works, Whitchurch,	
	Shropshire, SY13 1RS	
OPERATOR	Mr H Uebele,	
	Hyde Farm Nurseries, Marcham,	
	Nr Abingdon, Oxfordshire.	

Fig. 9.8   Hyde Farm Nurseries, Marcham, Oxfordshire

9.2.8 CASE STUDY 8 Stagshaw, Northumberland

This site is not connected to the national grid, and has used a 10kVA diesel generator to supply all electrical loads. Due to high electricity costs when running on part load and also a high level of maintenance (coking up), a 1kW Windharvester system was incorporated into the power system in February 1986.

The wind turbine system feeds a 340 Ah battery 24V. All power is drawn from the battery via a 2 kW inverter. When battery levels are low, an automatic relay brings in the diesel, providing continuous available power. The only device used on this small farm which cannot be run from the wind system is a 7kW shower unit, although in practice the diesel is used intermittently when demand exceeds available power in the battery.

EARLY REPORT   Watson, G R. Operational experience of hybrid wind-diesel systems. BWEA Small Wind Systems Conference, London, May 1982.

SITE   Aspect: Approx 900 ft above sea level
Open to prevailing winds, some shelter
150m to the North and East.

Mean Wind Speed approx 7 m/s

TURBINE   Rotor 3 bladed fixed pitch
2.8 metre diameter.
Upwind
Free yawing
Furledby flex in blades up to 30 mph and thereafter by tail swinging to bring rotor out of wind.

Rated output	1 kW
Rated wind speed	10.5 m/s
Cut in wind speed	4 m/s

TOWER   Guyed tubular steel lattice (galvanised)
Hub height   16.5 metres

ALTERNATOR   Permanent magnet, coupled direct to rotor

MANUFACTURER   Bergey Windpower Corporation
Norman, Oklahoma, USA

SUPPLIER   Northumbrian Energy Workshop Ltd.
Wind Energy Works, Acomb Industrial Estate,
Hexham, Northumberland, NE46 4SA

Fig. 9.9   Stagshaw, Northumberland

9.2.9 CASE STUDY 9 Hythe, Nr Folkestone,Kent

A Vortex 6000 (6kW) wind/turbine has been supplying the electricity requirements of an autonomous house in Hythe since 1980. Variable frequency 3-phase AC can be used directly for heating a 300 gallon water tank at up to 4 kW. Alternatively 3 x 1 kW night storage heaters can be supplied and in addition the wind turbine output can be rectified to provide battery charging. A substantial battery bank provides 220V DC which is inverted for AC applications such as the main frame computer which is owned by the operator (who is a computer engineer).

SITE	Aspect: open and flat
	Mean wind speed: 5 m/s
TURBINE	Rotor: 2 bladed, fixed pitch
	reinforced fibre-glass construction
	4.6m diameter
	Upwind
	Free yawing
	Rated output: 6 kW
	Rated wind speed: 12 m/s
	Cut-in wind speed: 3 m/s
TOWER	Steel lattice
	Height: 7.6m
ALTERNATOR	3 phase, synchronous generator,
	rectified for battery charging
MANUFACTURER	Natural Power Systems Ltd
	40 Sanderstead Road
	Croydon, Surrey CR2 OPA

Fig. 9.10   Hythe, Near Folkestone, Kent

9.2.10 CASE STUDY 10 Clapham, Lancashire

Since December 1983 a Windharvester 500 (500W) system has been providing the domestic electricity supply for this remote farmhouse on the border of Lancashire and N.Yorkshire. The house, which is privately owned, is not connected to the mains.

The wind system comprises a Sencenbaugh 500 HDS wind turbine and 680 A of battery storage, and supplies the dwelling with 12V DC which is used both or lighting and for low voltage DC appliances. To make the best use of the system the owner has geared her electricity demand toward this low voltage supply. Only for the intermittent use of larger power tools is a petrol/l.p.g. driven generator, which provides 240V AC, used

SITE   Aspect: open and undulating
                Mean wind speed: 7 m/s

TURBINE   Rotor: 3 bladed, fixed pitch
                         2m diameter
                         upwind
                         free yawing
                         furled by upward tilting of rotor

                Rated output   500w
                Rated wind speed   9.8 m/s
                Cut-in wind speed   3.8 m/s

TOWER   Steel lattice (galvanised)
                Hub height   15 m

ALTERNATOR   Direct drive with field wound rotor, rectified
                for battery charging and domestic supply

MANUFACTURER   Sencenbaugh Wind Electric Co.
(of wind   Paulo alto, California,
 turbine)   USA

SUPPLIER   Northumbrian Energy Workshop Ltd.
                Wind Energy Works, Acomb Industrial Estate,
                Acomb, Hexham, Northumberland NE46 4SA

Fig. 9.11   Clapham, Lancashire

9.2.11 CASE STUDY 11   Brunton Bank,Nr Hexham, Northumberland

Two Rutland WG910 50W wind turbines have been supplying electricity to this remote cottage since March 1984. The Windharvester 100 system which also includes 170Ah of battery storage was designed and installed by Northumbrian Energy Workshop Ltd. It provides a 12V DC supply for lighting and power points. AC is also available through a start—on—demand inverter which incorporates overload protection and low voltage cut out.

A standby petrol generator provides an AC power supply for intermittent larger AC loads, such as a washing machine. Manually switched charge links enable the battery store to be charged whilst the petrol generator is running.   A manually operated isolator facilitates the switch over from inverter to petrol generator.

Apart from a maintenance period the system has been operating continuously since its installation.

SITE	Aspect: open Mean wind speed: 6.5 m/s
TURBINES	Rotor: 6 bladed, fixed pitch 0.91m diameter Upwind Free yawing
	Rated output: 50W x 2 turbines Rated wind speed 9.8 m/s Cut-in speed 1.8 m/s
TOWER	Guyed tubular steel, (galvanised) Double mounting top bracket holds both machines in westerly direction Hub height 11.5 m/s
ALTERNATOR	Direct drive, permanent magnet type rectified output
MANUFACTURER (of wind turbines)	Marlec Engineering Co.Ltd. Unit 5, Pillings Rd.Industrial Estate, Oakham, Rutland LE15 6QF
SUPPLIER/ SYSTEM DESIGNER	Northumbrian Energy Workshop Ltd. Wind Energy Works, Acomb Industrial Estate, Acomb, Hexham, Northumberland NE46 4SA

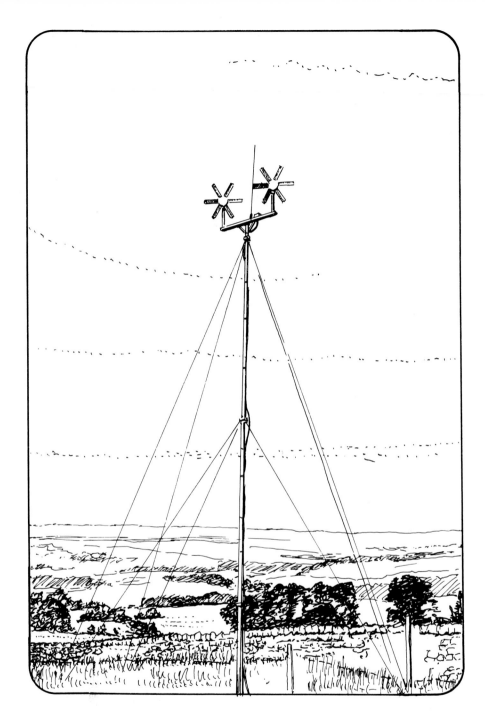

Fig. 9.12   Brunton Bank, Northumberland

123

9.2.12 CASE STUDY 12 Scalloway,Shetland

This was the first privately owned wind turbine connected to a utility grid after the 1983 Energy Act. This Act made it mandatory for the local electricity board to purchase excess electricity from the installation. The owner, Mr Arthur Williamson, purchased the Vestas 55kW/11kW twin generator machine as a Demonstration Project with support from the EEC and the Shetland Islands Council. It was installed in August 1985 at a site of tourist and residential chalets. The system is being monitored.

EARLY REPORT   Saluja G S, Robertson P and Stewart R S "A 55kW wind energy conversion system for a holiday chalet complex in the Shetland Isles", in 1986 Proc. of the BWEA, MEP Ltd. eds. M B Anderson and S J R Powles.

SITE   Awkward north west facing slope
Mean wind speed 8 m/s at 10m

TURBINE   Rotor 3 bladed, fixed pitch, upwind, horizontal axis
15 m diameter
Active yawing by motor drive
Rated output 55 kW max at 13 m/s
Cut in wind speed   4 m/s
Cut out wind speed   23 m/s

TOWER   Galvanised lattice
Tower height 22m

GENERATOR   Twin 55kW and 11kW induction generators

SUPPLIER   Mr T Buchanan,
Whinneyknowe House, Upper Polmaise,
Stirling, FK7 9PU

MANUFACTURER   Vestas Wind Systems A/S
DK-6940 Lem
Denmark

OPERATOR   Mr L A Williamson, Scalloway, Shetland

Fig. 9.13   Scalloway, Shetland

OPERATING PERFORMANCE

An average annual output of 153,000 kWh is expected from the measurements
made since installation.

APPENDIX 1 : BIBLIOGRAPHY

A1-1: Books

1. British Wind Energy Association (1982) "Wind energy for the Eighties". Peter Peregrinus Ltd., Stevenage, UK.

2. Freris, L (Ed) (in preparation 1987) "Principles of Wind Energy Conversion Systems" Prentice Hall, London.

3. Golding, E W (1976) "The Generation of Electricity by Wind Power" E & F N Spon, London.

4. Twidell, J and Weir, A D (1986) "Renewable Energy Resources", Chapter 9. E and F N Spon, London.

5. Warne, D F (1983) "Wind Power Equipment" E and F N Spon, London.

6. Wegley, H L, Ramsdell J V, Orgill M M and Drake R L (1980). "A siting handbook for Small Wind Energy Conversion Systems" WindBooks, Pacific Northwest Laboratory, Richland, Washington 99352.

7. World Meteorological Organisation (1981) "Meteorological Aspects of the Utilisation of Wind as an Energy Source". Technical Note No.175, World Meteorological Organisation, Geneva, Switzerland.

8. "Kemps Engineering Yearbook" (updated annually) Morgan Grampian Ltd. London.

A1-2: Journals and Periodicals (* Journals include some articles on wind turbines)

1. *Modern Power Systems incorporating Energy International. Published monthly by United Trade Press Ltd.
   P O Box 54
   Basildon
   Essex SS15 6SS

2. *Renewable Energy News: available from
   Energy Technology Support Unit
   Building 156
   AERE Harwell
   Didcot
   Oxon OX11 0RA

3. *Solar Energy: International journal for Scientists, Engineers,
    and Technologists in solar energy and its application
    Published monthly by European Publishing Corporation Ltd.
                           84-86 Mahide Road
                           Cookick
                           Dublin 5
                           Republic of Ireland
        Subscription Office: Headington Hill Hall
                           Oxford, OX30 BW1

4. Wind Engineering: The official journal of the European Wind
    Energy Association and the British Wind Energy Association.
    Published quarterly Multi-Science Publishing Co.Ltd.
                           42/45 New Broad Street
                           London EC2M 1QX

5. Wind Engineering and Industrial Aerodynamics.
        Published 9 times per annum by:
                           Elsevier: Science Publishers B.V.
                           P O Box 211, 100 EE Amsterdam
                           Netherlands

6. Wind Industry News Digest
        Published twice monthly by:
                           Alternative sources of energy
                           107 So.Central Avenue
                           Milaca, MN 56353, USA

7. Wind Power Monthly.
        Published monthly by:Farlaget Vistoft ApS
                           Vrinners Hoved
                           DK-8420 Knebel
                           Denmark

8. Windirections:
        4 issues per annum.
        Published by: British Wind Energy Association
                           4 Hamilton Place
                           London W1V OBQ

Al-3 Government Publications

1. Department of Energy, Electricity Division. Energy Act 1983.
    Private Electricity Supply - Informal Guide.

2. Electricity Council, Engineering Recommendation G59.
    Recommendations for the connection of Private Generating
    Plant to the Electricity Boards' Distribution Systems.

3. Electricity Council. Safety Branch document ECSB/84/03-28.Guidance
    Notes on the Safety Implications of the Energy Act, 1983.

4. Electricity Supply Regulations, 1937. Explanatory Notes on the
    Electricity Supply Regulations 1937, prepared by the Electricity
    Commissioners. Available from Her Majesty's Stationery Office.

5. Factories Act 1961. Memorandum on the Electricity Regulations.
    This memorandum explains the effect of Electricity (Factories
    Act) Special Regulations, 1908 and 1944. Available from HMSO.

6. Health and Safety Commission. Various leaflets giving advice
    on the provisions of the Health and Safety at Work etc.Act,1974.
    Available from Her Majesty's Stationery Office.

7. Health and Safety Commission. Guidance Notes G58. Articles and
    Substances for Use at Work. Guidance for designers,
    manufacturers, importers, suppliers, erectors and installers.
    Available from Her Majesty's Stationery Office.

8. Health and Safety Executive. A Guide to Agricultural Legislation
    HSR(2). Available from Her Majesty's Stationery Office.

9. Health and Safety Executive, Guidance Note PM53. Emergency Private
    Generation: Electricity Safety. Available from HMSO.

10. The Electricity(Private Generating Stations and Requests by
    Private Generators and Supplies) Regulation 1984 no.136.

11. House of Commons Committee on the Problem of Noise (1963).
    Reference Command 2056, Volume 22, page 657 (Hansard).

Al-4 Articles

1. "Man on the Job", leaflets, Cement and Concrete Association,
    Wexham Springs, Slough SL3 6PL.

2. Soderquist (1982) "Swedish WTG : the noise problem",
    Proc. 4th International Symposium on Wind Energy Systems.
    Aeronautical Research Institute of Sweden, Stockholm.

APPENDIX 2a : USEFUL ADDRESSES

Association of British Insurers, Aldermary House, Queen Street,
    London EC4N 1TU

British Standards Institution, 2 Park Street, London W1A 2BS

British Telecom, 2-12 Gresham Street, London EC2 7AG

British Wind Energy Association, 4 Hamilton Place, London W1V OBQ

Department of Energy, Electricity Division, Thames House South,
    Millbank, London SW1P 4QJ

Department of the Environment, 2 Marsham Street, London SW1P 3EB

Department of Trade and Industry, 1 Victoria Street, London SW1H OET

Department of Trade and Industry, Radio Investigation Service,
    Directorate of Radio Technology, Waterloo Bridge House, Waterloo
    Road, London SE1 8UA

Electricity Boards - Channel Islands
    Alderney Electricity Ltd., 40 Victoria Street, St Anne's, Alderney
    Guernsey Electricity Board, PO Box 4, Electricity House,
       North Side, Vale
    Jersey Electricity Co.Ltd., PO Box 45, Queens Road, St Helier

 Electricity Boards - England and Wales:

    Central   : Central Electricity Generating Board, Sudbury
                          House, 15 Newgate Street, London EC1A 7AU
    Eastern   : PO Box 40, Wherstead, Ipswich IP9 2AO
    East Midlands   : PO Box 4, North PDO, 398 Coppice Rd, Arnold,
                          Nottingham NG5 7HX
    London   : Templar House, 81-87 High Holborn,London WC1V 6NU
    Merseyside &
     North Wales   : Sealand Road, Chester CH1 4LR
    Midlands   : Mucklow Hill, Halesowen, West Midlands B62 8BP
    North Eastern   : Carliol House, Newcastle upon Tyne NE99 ISE
    North Western   : Cheetwood Road, Manchester M8 8BA
    South Eastern   : Grand Avenue, Hove, East Sussex BN3 2LS
    South Wales   : St Mellons, Cardiff CF3 9XW
    South Western   : Electricity House, Colston Avenue, Bristol BS1 4TS
    Southern   : Southern Electricity House, Littlewick Green,
                          Maidenhead, Berkshire SL6 3QB
    Yorkshire   : Scarcroft, Leeds LS14 3HS

Electricity Boards - Isle of Man

Douglas Corporation, Ridgway Street, Douglas, Isle of Man
Isle of Man Electricity Board, Harcroft, Douglas, Isle of Man

Electricity Boards - Northern Ireland

Northern Ireland Electricity Service, 120 Malone Road,
Belfast BT9 5MT

Electricity Boards - Scotland

North of Scotland Hydro-Electric Board, 16 Rothesay Terrace,
Edinburgh EH3 7SE
South of Scotland Electricity Board, Cathcart House, Spean St,
Glasgow G44 4BE

Electricity Council, 30 Millbank, London SW1P 4RD

Engineering Insurance Companies:

British Engine Insurance Ltd, Longridge House, Manchester M60 4DT

Eagle Star Group Engineering Insurance Ltd., 54 Hagley Rd,
Edgbaston, Birmingham B16 8QP

National Vulcan Engineering Insurance Group Ltd, St Mary's Parsonage,
Manchester M60 9AP

Scottish Boiler and General Insurance Group Ltd. Windsor House,
250 St Vincent St, Glasgow G2 5UT

Health and Safety Executive (Policy), Baynards House, 1 Chepstow
Place, London W2 4TF

Health and Safety Executive (Factory Inspectorate, Medical Division,
etc.), St.Hughs House, Stanley Precinct, Bottle, Merseyside L20 3QA

Her Majesty's Stationery Office - Government Bookshops:

49 High Holborn, London WC1V 6HB
13a Castle Street, Edinburgh EH2 3AR
41 The Hayes, Cardiff CF1 1JW
Brazenose St, Manchester M60 8AS
Southey House, Wine Street, Bristol BS1 2BQ
258 Broad Street, Birmingham B1 2HE
80 Chichester St, Belfast BT1 4JY

Institution of Electrical Engineers, Savoy Place, London WC2R OBL

National Engineering Laboratory, East Kilbride, Glasgow G75 OQU

Scottish Economic Planning Department, New St.Andrews House,
    Edinburgh EH1   3TA

U.K. National Wind Turbine Centre, National Engineering Laboratory,
    East Kilbride, Glasgow G75 OQU   (03552-20222)
    (Test Site, Myres Hill, Eaglesham : no postal address)

APPENDIX 2b **List of** U.K. **Meteorological** Wind Recording Stations
(latitudes North; longitudes West; unless marked
E for East)

Name	Lat.	Long	Name	Lat.	Long
Lerwick	60.08	01.11	Isle of Grain	51.27	00.43E
Sumburgh	59.53	01.18	Dungeness	50.55	00.58E
Kirkwall	58.57	02.54	Dover	51.07	01.20E
Dounreay (high)	58.35	03.44	Manston	51.21	01.21E
Dounreay (low)	58.35	03.44	Thorney Island	50.49	00.50
Wick	58.27	03.05	Brighton Marina	50.49	00.06
Shin	57.57	04.25	Hurn	50.47	01.50
Stornoway	58.13	06.20	Middle Wallop	51.09	01.34
Duirinish	57.19	05.41	Calshot	50.49	01.18
Benbecula	57.28	07.22	Lee-on-Solent	50.48	01.13
Corpach	56.50	05.09	Farnborough RAE	51.17	00.45
Fort Augustus	57.08	04.43	Lyneham	51.30	01.59
Dalcross	57.33	04.03	Larkhill	51.12	01.48
Cairngorm	57.08	03.39	Boscombe Down	51.09	01.45
Kinross	57.39	03.34	Porton	51.07	01.42
Lossiemouth	57.43	03.20	Tiree	56.30	06.53
Dyce	57.12	02.12	Machrihanish	55.26	05.42
Fraserburgh	57.41	02.20	Dunstaffnage	56.28	05.26
Mylnefield	56.27	03.04	Inchterf	55.57	04.07
Bell Rock	56.26	02.24	Greenock Port	55.75	04.46
Rannoch	56.41	04.35	Paisley	55.50	04.25
Tummel Bridge	56.42	04.01	Abbotsinch	55.52	04.26
Leuchars	56.22	02.53	East Kilbride (low)	55.45	04.10
Forth Road Bridge	56.00	03.24	Lowther Hill	55.23	03.45
Turnhouse	55.57	03.21	Ardrossan	55.38	04.49
Edinburgh Obser'ry	55.55	03.11	Prestwick	55.30	04.35
Lynemouth	55.12	01.32	Dumfries	55.03	03.39
Blyth	55.08	01.30	Eskdalemuir	55.19	03.12
Boulmer	55.25	01.36	West Freugh	54.51	04.57
Durham	54.46	01.35	Ronaldsway	54.05	04.38
South Shields	55.00	01.26	Snaefell	54.18	04.28
Leeming	54.18	01.32	Point of Ayre	54.25	04.22
Linton-on-Ouse	54.03	01.15	Sellafield (low)	54.25	03.31
South Gare	54.38	01.08	Sellafield (high)	54.25	03.31
Silpho Moor	54.20	00.31	Carlisle	54.56	02.57
Scampton	53.19	00.33	Gt Dunfell	54.41	02.27
Waddington	53.10	00.31	Moor House	54.41	02.23
Cranwell	53.02	00.29	Squires Gate	53.46	03.02
Coningsby	53.05	00.10	Aigburth	53.22	02.55

continued next page   continued next page

Name	Lat.	Long.	Name	Lat.	Long.
Binbrook	53.27	00.12	Fleetwood	53.56	03.01
Marham	52.39	00.33E	Manchester WC	53.29	02.15
Coltishall	52.46	01.22E	Ringway	53.21	02.16
Gorleston	52.35	01.43E	Valley	53.15	04.32
Honington	52.20	00.47E	Trawsfynydd (low)	52.55	03.57
Wattisham	52.07	00.58E	Bala	52.54	03.35
Wyton	52.21	00.07	Aberporth	52.08	04.34
Bedford	52.13	00.29	Milford Haven	51.42	05.03
Cardington	52.06	00.25	Brawdy	51.53	05.08
Garston	51.42	00.23	Port Talbot	51.34	03.45
Stansted	51.53	00.14E	Rhoose	51.24	03.21
Shoeburyness	51.32	00.49E	Cilfynydd	51.38	03.18
Wilsden	53.49	01.52	Yeovilton	51.00	02.38
Sheffield Univ.	53.23	01.29	Portland Bill	50.31	02.27
High Bradfield	53.26	01.35	Plymouth	50.21	04.07
Nottingham Met.Cen.	53.00	01.15	Burrington	50.46	03.59
Finningley	53.29	01.00	Exeter	50.44	03.25
Witte Ring	52.37	00.28	Scilly	49.56	06.18
Edgbaston	52.28	01.56	Gwennap Head	50.02	05.40
Elmdon	52.27	01.44	Culdrose	50.05	05.15
Brize Norton	51.45	01.35	Lizard	49.57	05.12
Oxford	51.46	01.16	St Mawgan	50.26	05.00
Benson	51.37	01.05	Coleraine Univ.	55.09	06.40
Keele	53.00	02.17	Aldergrove	54.39	06.13
Shawbury	52.48	02.40	Ballypatrick		
Preston Wynne	52.07	02.30	Forest	55.10	06.09
Avonmouth	51.30	02.43	Belfast Harbour	54.38	05.53
Innsworth	51.53	02.12	Kilkeel	54.03	05.59
Whitehall Gdns.	51.27	00.10	Killough	54.14	05.37
London WC	51.31	00.07	Orlock Head	54.40	05.35
Post Office Tower	51.31	00.09	Carrigans	54.40	07.19
Blackwall	51.31	00.05E	Castel Archdale For	54.28	07.42
Heathrow	51.28	00.28	Guersey A/P	49.26	02.36
Kew	51.28	00.19	Jersey A/P	49.12	02.11
Gatwick	51.09	00.11			

Contents

A3.1 INTRODUCTION

The U.K. Energy Act 1983 clearly establishes the rights of private suppliers to market electricity directly to their own customers and indirectly to the public through the nationalised Electricity Board networks.   Supply from aerogenerators was specifically mentioned during the legislative procedures.   The technical and economic factors are clear for large aerogenerators established predominantly for the export of power to the network but extremely uncertain for small aerogenerators exporting intermittent power.

The historical development of electricity supplies has been closely related to technical capabilities, public safety and amenity, standards of supply, government regulations and commercial opportunity.   In the U.K. the nationalisation of electricity supply, especially through the 1943 Act that established the North of Scotland Hydro-Electric Board and the 1947 Act, was of great significance.   However these Acts must be seen in the context of a continuous series of government regulations extending for a hundred years of community electricity supply.   All countries show similarities of legislation, much related to technical developments and supply economics.   The O.P.E.C. oil price increases of 1973 have initiated in recent years much innovative thinking aimed at diversifying generation and at encouraging alternative methods of supply.

The U.K. Energy Act must be seen in this historical international framework.   Its importance is that government has affirmed its support of private generation of electrical power that may be marketed directly to other private organisations or indirectly to the public via the nationalised Electricity Boards (utilities).   The Boards are vital intermediaries in this process, since they must ensure technical adequacy and an economic balance so that their own general customers have electricity neither increased or decreased in price by any one marginal arrangement with a private supplier.   In these aspects the Energy Act is not unlike other new legislation in North America and Europe (Lovasz 1984).   However the interpretation of the legislation may differ considerably in the U.K. in view of opinions about alternative sources of power and the very large over-capacity in some Electricity Generating Board networks.

The U.K. Energy Act does however have one component that is attracting much interest from other countries. A private supplier has a right to 'hire' the use of an Electricity Board's network for marketing power to his own customers. This provision immediately increses both the range of interest of private suppliers and the technical difficulties for the Board's operation.

A3.2 THE ENERGY ACT AND ITS REGULATIONS
The legislation has advanced by three important stages:

a   Debate in Parliament as reported in Hansard (Lawson 1982);

b   The Energy Act, that expresses the purpose of the legislation, outlines the legal requirements on Boards and private suppliers, and establishes methods to settle disputes; and

c   The subsequent regulations.

Interaction between the private suppliers and the appropriate Electricity Board is affected by each Board's tariff structure and the regulations (national and local technical codes of practice). the legislation has been reviewed by Evans (1984). Walker (1984) has described the regulations and codes of practice, although at the time of writing many of these were under review by the relevant authorities. He particularly refers to technical factors that could affect aerogenerators.

For the private owner of an aerogenerator the important new factor is that a single and specific Act of Parliament lays down his rights for connection to the Board's grid, for selling power and for using the transmission network. The Boards must deal positively with applications for such arrangements, and there is an appeal structure in case of disagreement. A further benefit for all parties is that tariff structures and technical requirements are published by each Board in a much more purposeful manner than before the Act. In addition Boards may offer special terms and arrangements for individual applictions or, more likely, certain classes of systems (e.g. small aerogenerators). The North of Scotland Hydro-Electric Board has already published its special terms for connecting small generators in response to difficulties met by private suppliers (N.S.H.E.B. 1984).

At the present time a number of intending private suppliers and the Boards are carefully exploring the implications of the Act. As in all such cases where local factors are of great significance, an intending supplier is well advised to put his ideas to the local Board transmission department as early as possible for discussion, and also to obtain case histories of other installations.

135

A3.3 TARIFF STRUCTURES

Tariff structures of the many U.K. Boards have been interpreted by Tolley (1983 and 1984) in several publications for the general private supplier of electricity. It is probable that the largest possible component of such supply is foreseen from combined heat and power systems in industry. The Department of Energy has put great emphasis on encouraging such schemes in the recent energy efficiency campaign, and obviously interconnection is of both technical and economic importance. Shaw & Bossanyi (1984) have analysed the English and Welsh Boards' tariff structures in considerable detail, and expressly related these to small aerogenerators connected to the Boards' low voltage transmission lines. A further publication (Clare et al. 1984) includes the associated, but different, question of assessment of aerogenerators for local rating taxation. It is noteworthy that the Energy Act itself appears to have neglected the differences in local taxation rates between the Electricity Boards and private suppliers.

The general principles for the tariffs are in common with all Boards. Essentially the private supplier must pay for all the benefits obtained by using the Board's facilities. Anything otherwise would imply that extra cost was being charged to the other customers of the Board. In turn the Board must pay for the benefits it receives from being able to market the private supplier's electricity at the Board's normal sales price. An important benefit for the Board is offsetting marginal increases in the construction of extra generating capacity, and the decrease of its own operating costs. These latter factors are important in diesel generating systems and wherever new plant is needed.

A3.3.1 Between the Board and the private supplier (selling-only)
The components for these charges are:

a   Initial connection charge (£ per specific installation).

b   Metering for exported power.

c   Fixed annual charge per supply point, which may not be charged for private sales only (£ per annum).

d   Reactive power charge to compensate for changes in local characteristics (£ per kVArh per month of maximum demand). N.B. a seller of power may be a net importer of reactive power, e.g. for induction generators.

e   Export capacity charge to cover the Board's extra costs for transmission (£ per kVA per month).

f   Unit charges (pence per kWh). The private supplier receives money for the kWh units put into the grid. The unit price varies markedly between summer and winter, and between evening and daytime. Shaw and Bossanyi (1984) give the range between 1.33 and 5.52 pence per kWh. It is therefore of the greatest importance that the private supplier tries to optimise the timing of exported excess power by arranging his own private load in the best manner for his own advantage.

A3.3.2 Between the Board and the private supplier (who both sells and buys)

It appears to be a matter of individual negotiation whether there is a double accounting or other discrepancy between charges if the private supplier also purchases power.

a Initial connection charge (£ per specific installation)

b Metering for exported power. Metering of imported power is not expected as a specified charge.

c Fixed annual charge per supply point (£ per annum).

d Reactive power charge (£ per kVArh per month of peak power demand). The detailed charges are usually different from A3.3.1(d) above.

e Availability of capacity charge (£ per kVA per month of peak power demand). This charge is to pay for the generating capacity that the Board has to establish in case the private system requires power. The charge is likely to be made irrespective of whether this peak power is in fact demanded, in which case the peak power is the authorised capacity. It is this charge that is most contested if the private system also sells, since the private supplier may consider there is double accounting unless these fixed charges are somehow rebated against the net exchange of power. In practice it is to everyone's advantage that the authorised maximum capacity is made as low as possible. Load management is an obvious method of lowering maximum demand.

f Demand charges (£ per kVA of peak power actually consumed). This charge usually varies between winter and summer months. Load management can greatly reduce this charge.

g Unit charges (pence per kWh). This charge will vary for commercial installations by month through the year, and by period in twenty four hours. Shaw and Bossanyi (1984) give the range between 1.52 and 9.97 pence per kWh with English area Electricity Boards. Load management is clearly of great benefit. Indeed there are some times when a Board's buying-in price is greater than its selling price, although this is unusual. In general the Boards appear to sell power units for at least 20% to 50% more than they buy-in at any specific time. However this marginal 'handling charge' may increase greatly if time-specific metering equipment is not installed for a private supplier.

137

A3.3.3. 'Hire' of the Board's transmission lines

An important principle of the tariff structure is that any other consumer supplied by the private generator becomes a customer of the private supplier and not the Board. Thus the Board deals only with one agent who negotiates independently with his own set of customers.

The Boards publish tariffs for the use of their transmission systems. The rates depend on the points of injection and extraction, the number of transformations and the authorised capacity. It is obviously to everyone's benefit that the transmission systems operate at steady and peak loads, so load management is necessary. The private supplier will be expected to pay for metering, which is necessarily complex if extraction is to be always equal to injection.

A3.4 CONCLUSIONS

The direct substitution of wind generated power for conventional grid supplied electricity may not be obviously economically attractive in the U.K. at present. This conclusion is reached for grid connected aerogenerators of capacity less than about 100 kW under terms operating before and after the 1983 Energy Act; assuming that no institutional grants or subsidies are specifically available for wind machines, assuming no load management control by the private consumer or supplier, and assuming the unfair difference of local rating taxation continues between the public and private suppliers.

However, there are several factors that have not yet been fully explored in the U.K, each of which could significantly improve the competitiveness of aerogeneration. These include:

a   Load management. Systems with a substantial amount of heating are very attractive for wind power. Load control equipment responding to peak currents, and time of day and year will be important if power is exchanged with the grid. It is in everyone's interest for private suppliers to use a maximum of their own power and to minimise peak power exchanges with the grid. Microprocessor controlled switching is ideally suited for this form of energy management.

b   Diesel powered grids, which should provide particularly attractive export prices for the private supplier. This in itself would provide sufficient commercial incentive for significant U.K. involvement in wind farms and agriculturally related machines. Examples include the Isles of Scilly, the Shetlands, the Western Isles, the Isle of Man and the Channel Islands.

c Wind/hydro linkage. One concept missing from the Energy Act is any provision for the storage of privately generated power. Thus intermittent supplies of wind generated power can be effectively stored if reservoir linked hydro power is in the grid system. In this way wind generated energy can be made available at times of peak supply.

These suggestions arise from the realisation that significant benefits only arise from recognising the distinctive attributes of wind power, and the need for sophisticated analysis for particular systems. Such analysis must include all local energy demands with an understanding of their priority. Modern technology is now available to allow this.

REFERENCES

Evans J H (1984). "The Energy Act - statutory aspects", in Twidell J W (1984), "The Energy Act and other institutional aspects of wind power generation", Proc. BWEA Day Meeting, University of Strathclyde, Glasgow.

Clare R, Shaw T L and Bossanyi E (1984). "The economics and local taxation position of privately owned wind turbines operated by consumers to the UK national electrical network". In Musgrove (1984) (ed) Proc. of the 1984 BWEA Annual Conference, C.U.P., Cambridge.

Holding N (1984). "Small aerogenerators on farms", in Twidell (1984) "The Energy Act etc.", ibid.

H.M.S.O. (1983). Hydro-Electric Development (Scotland) Act, H.M.S.O., London.

H.M.S.O. (1947). Electricity Act, 1947: ch.54 H.M.S.O.,London.

H.M.S.O. (1983). Energy Act 1983 ch.25, H.M.S.O., London.

Jongens P (1984). "Tariff implications of the Energy Act", in Twidell J W (1984) "The Energy Act etc.", ibid.

Lawson Rt.Hon.Nigel (1982). Parliamentary debates (Hansard) 24th Nov. 1982, 32 No.16, columns 868, 869, 871.

Lovasz J, (1984). Report for the EEC. Didier and Associates, Rue Vergote 11, B - 1040, Brussels. (Electricity tariffs in the EEC).

Musgrove P (1984) (ed). Proceedings of the 1984 BWEA Annual Conference, C.U.P., Cambridge.

N.S.H.E.B. (1984). "Special terms for supply of electricity to small generators connected before 1 Jan.1983", Section A13, schedule 2.1, North of Scotland Hydro-Electric Board,Edinburgh.

Shaw T L, and Bossanyi E (1984). "The economics of small privately-owned wind turbines operated by consumers connected to the U.K. National Network", in Twidell J W (1984), "The Energy Act etc", ibid.

Sinclair B A, Somerville W M and Stevenson W G (1984). "Wind-power generation on Fair Isle". Proceedings of Energy for Rural and Island Communities" III, Twidell J W, Riddoch F, and Grainger, W, (eds), Pergamon Press, Oxford.

Sinclair B A (1984). "Accounts of the Fair Isle Electricity System", (private communication).

Stevenson W G and Somerville W M (1983). "The Fair Isle Wind Power System". Proceedings BWEA Annual Conference, (ed. P Musgrove), C.U.P., Cambridge.

Tolley D (1983, 1984). Various publications. Tariff Application Group. Commercial Department, Electricity Council, London.

Twidell J W (1984) (ed). "The Energy Act and other institutional aspects of wind power generation", Proceedings of BWEA Day Meeting, University of Strathclyde, Glasgow.

Twidell J W (1984). "Wind Power and the UK Energy Act (1983)" in Proc. 6th BWEA Conference, (ed. P Musgrove), C.U.P., Cambridge.

Walker J F (1984). "Some technical implications of the Energy Act", in Twidell J W (1984) "The Energy Act etc", ibid.

APPENDIX 4: EXAMPLE OF COSTING A VERY SMALL MACHINE

A4.1 The outright purchase

A wind turbine of 5 kW rated capacity is installed with the aim of providing 15,000 kWh of electricity in a period of one year (8760 hours). The turbine is therefore expected to operate at a capacity factor of 34%, i.e. it produces on average 1.7 kW throughout the year at this particular site. It is assumed in this example that the SWECS is purchased to substitute for energy from a high cost system, such as diesel generated electricity. We assume the displaced energy had a cost of 10p/kWh. Such a high cost estimate is not excessive if all the costs appropriate to the displaced power, and a measure of efficiency, are included.

In this example it is assumed that improvement grants provide a proportion of the purchase price. In the case of leased equipment, the owner who received the grant will not be its operator, but the leasing company who passes on the benefit of lower capital cost in the monthly or quarterly leasing charges.

The ex-works price of the wind turbine is estimated for the purposes of this example at ₤5000, that is ₤1000 per kW capacity. The delivery, foundation, and erection costs are taken to be 25% of the ex-works price, to give an installed price of ₤6250. This is therefore a high cost example, since lower prices can be expected in the future.

Insurance premiums have been assumed to be 2.5% per annum of first cost. The local authority rates are taken as 0.8% per annum of first cost. The 0 & M costs rise progressively over the life of the equipment. No cost has been included for maintenance contracts.

The turbine depreciation charge has been calculated on the basis discussed where the 'lifed parts' are considered as 25% of the turbine first cost, and are depreciated at a different rate to the rest of the structure. Grants may be paid on the agreed overall cost price of the wind turbine and therefore this will not be the same sum as that used for depreciation. There are three values for accounting purposes as below:

** The overall cost of the wind turbine on which grants are paid.

** The cost of the wind turbine, less the value of the lifed components depreciated at a rate appropriate to the life of the turbine.

** The individual lifed assemblies. These are depreciated at rates determined by their lifetime.

A4.2 Depreciation calculation

Ex works price:

Tower (not lifed)	500
Turbine machinery (not lifed)	3250
Turbine machinery (lifed)	1250
TOTAL EX-WORKS COST	£5000
Installation costs including fees and foundations (not lifed)	£1250
TOTAL CAPITAL	£6250

The non-lifed components, the tower, and the installation costs will be depreciated over the 20 year life of the wind turbine at the rate of 5% per annum. The lifed components are to be replaced after 10 years' service and will be depreciated at 10% per annum. Thus the depreciation charge for the first year is:

Non lifed components	10% on 5000	= 500
Lifed components	20% on 1250	= 250
	TOTAL	£750

A4.3 Accountancy rate of return

The accountancy rate of return method expresses the average estimated yearly net cash inflows as a percentage of the net investment outlay. As the depreciation may be covered by the yearly net cash inflow, a formula may be expressed:

$$R = (C - D)/I$$

where R = the accounting rate of return on capital

C = average yearly net inflow

D = depreciation charge

I = net investment outlay

A4.4 Outright purchase without grant aid

Year	1	2	3	4	5	Sum of 5 years:£
Insurance	156	156	156	156	156	780
Local rates	100	100	100	100	100	500
O & M	25	30	35	40	50	180
TOTAL OUT-GOINGS	281	286	291	296	306	1460
Income (displaced electricity at 10p per kWh	1500	1500	1500	1500	1500	7500
Net inflow C	1219	1214	1209	1204	1194	6040
Depreciation D decreasing balance method	750	600	480	384	307	2521
C - D	469	614	729	820	887	3519
Investment	6250	6250	6250	6250	6250	6250
Accounting rate of return on capital R = (C-D)/I (%)	8	10	12	13	14	

The total cost over 5 years is the sum of the depreciation (£2521) and the outgoings (£1460). This equals £3981. The energy produced over 5 years is 75,000 kWh at a unit cost of 5.3 p/kWh.

A4.5 Outright purchase with grant aid

Grants (say of 18%) towards first cost will reduce the value of the wind turbine to be depreciated and therefore increase the return on capital. Grants do not have any influence on the operating costs, rates or insurance premiums associated with the wind turbine. They are paid only once in respect of each item of equipment.

The Institute of Chartered Accountants' Standard Accounting Practice (M4) 2.104 recommends the creation of a deferment reserve of the grant monies paid reducing this balance at the same rate as the depreciated asset each year. The debited cash is credited to the credit reserve in the profit and loss account. The depreciation charge then reflects the cost of the asset throughout its life. Grants may vary in amount or cease altogether, but treated this way, the cost of the asset will not be distorted in the profit and loss account. The table below gives the analysis.

Year	1	2	3	4	5
Depreciation	750	600	480	384	307
Grant credit	56	53	50	48	46
Cash flow in	1275	1267	1259	1252	1240
Return on capital (%)	8	11	12	14	15

The immediate effect of the grant is to increase the return on capital employed. The unit cost of electricity generated also falls from 5.3p to 5.0p (a reduction of 5.7%).

A4.6 Credit purchase

It is common practice to borrow a proportion of the purchase costs from a finance company so that not all the capital cost of the project need be found at one time. The finance company must be paid for this facility so the total expenditure will rise to provide for these extra charges.

It is imprudent to buy a wind turbine on borrowed money until clear evidence is provided by the vendor that the turbine will produce the power expected in the chosen locality. Experience of users elsewhere should give a high level of confidence that the quality and reliability of the equipment will meet the duty expected of it.

If the same example is taken again, the cost of depreciation will be less. This is because the cost reflects the depreciation on the down payment made at the time of purchase. In the following example the down payment is 32% of the installed cost of £6250, i.e. £2000.

The hire purchase is charged at the rate of 10% on the outstanding balance of £4250 over the five years of the loan period (i.e. £425 of interest, plus £850 of repayment per year, total financial charge £1275 per year).

Year	1	2	3	4	5	Sum of 5 years
Insurance	156	156	156	156	156	780
Rates	50	50	50	50	50	250
O & M	25	37	56	85	125	328
Finance Cost	1275	1275	1275	1275	1275	6375
TOTAL OUTGOINGS	1506	1518	1537	1566	1606	7733
Depreciation D	200	160	128	102	82	672
Income (displaced electricity at 10p/kWh	1500	1500	1500	1500	1500	7500
Net inflow C	−6	−18	−37	−66	−106	−233
C − D	−206	−178	−165	−168	−188	−905

The total outgoings over the first five years for the wind turbine are
£7733.   This gives a unit cost of electricity for the 75,000 units
produced of 10.3p/kWh.   In the first example the unit cost was 5.3p/kWh.
So the effect of the finance charges has been to double the cost of
electricity from the wind generator for the first five years when the
cost of the wind turbine is recovered.   This cost drops dramatically
once the hire purchase has been repaid.

The hire purchase charges are analogous to a high depreciation charge
with the added cost of the interest on the borrowed money.   Thereafter
the cost of power is reduced and the cash flow goes positive.   This
suggests a pay back period of 5.25 years for the outright purchase, and
7.5 years for the credit purchase.

A4.7   EFFECT OF INFLATION

So far in this discussion the examples have been worked
out at a constant purchasing power of money.   However money values are
more likely to be eroded by inflation.   This drift in money values makes
it difficult to compare investment opportunities for the basis of
comparison is forever changing.   The technique of the net present value
of the asset has been developed to overcome this difficulty.

The discounted cash flow (DCF) method is now widely used, for it
recognises that the value of money is subjected to a time reference.
The DCF compares the net cash flows that accrue over the life of the
investment at their present day value, with the funds presently to be
invested.   It is thus possible to compare the rate of return in a
realistic manner.   Tables of DCF factors for predetermined rates of
interest are widely published. Taking this concept a stage further, the
annual charge rate, C, is defined as:

$$C = \frac{r}{1 - (1 + r)^n}$$

where r = the real rate of return allowing for inflation

n = machine lifetime in years

Table A4.1 shows simple pay back periods in years that relate to the
first year savings and the rate of increase in fuel costs.

In periods of high inflation especially. it is best to purchase
equipment that is well made and robust so that it can achieve a useful
life of 15-20 years.   Realistic charge rates of 20% are necessary and
can be achieved in practice.

Table A4.1. Years to simple payback

First year energy saving as % of system cost	Energy price inflation rate, %/year							
	5	6	7	8	9	10	11	12
2	26	24	23	21	20	19	18	18
4	17	16	15	15	14	14	13	13
6	13	12	12	12	11	11	10	10
8	10	10	10	10	9	9	9	9
10	9	9	8	8	8	8	8	7
12	8	7	7	7	7	7	7	7
14	7	7	6	6	6	6	6	6
16	6	6	6	6	6	6	6	5
18	6	5	5	5	5	5	5	5
20	5	5	5	5	5	5	5	5
22	5	5	5	5	4	4	4	4
24	4	4	4	4	4	4	4	4
26	4	4	4	4	4	4	4	4
28	4	4	4	4	4	4	4	3
30	3	3	3	3	3	3	3	3
40	3	3	3	3	3	3	3	3

A4.8 EXAMPLE OF DISCOUNTED CASH FLOW EVALUATION

The present value of a future expenditure is the money that should be invested now, at a certain interest rate, to become equal to the expenditure in the future. For example:

5% discount rate

Year	Factor	Cash inflow	Present value
1	0.9524	1219	1161
2	0.9070	1214	1101
3	0.8638	1209	1044
4	0.8227	1204	991
5	0.7835	1194	935
6	0.7462	1184	884
7	0.7107	1173	834

Present value of net inflows (sum for 7 years)   = £6908
Cost of investment outlay (capital cost)   = £6250

NET PRESENT VALUE OF PROJECT = £ 658

As the present value of the inflows over the period is in excess of the present outflows, the investment should go ahead if the seven year period meets the investment criteria laid down.

A4.9 VALUE ADDED TAX

In the UK value added tax (VAT) is charged on wind turbines. The tax is also charged on consultants fees and on the foundation works if carried out by a contractor.

For running costs, insurance premiums and local rates are free of VAT, however all maintenance and other operational charges will carry this tax. In this chapter, VAT has not been included because many operators will be able to claim this tax back.

A4.10 ECONOMIC CONSIDERATIONS OF GRID CONNECTED MACHINES

An owner will install a SWECS either to produce power where an adequate supply was not available previously, or to substitute for a conventional fuel. In the latter case it is possible to make a fairly accurate economic comparison of the choices. In the former case the owner will just have to decide on how much money is to be spent for the supply itself.

As illustrations of the many methods of economic assessment, consider the following examples.

C = initial installed cost (£)
B = interest rate of borrowing money at a flat rate over the period involved (interest only) (%/y)
I_1 = general rate of inflation (%/y)
I_2 = inflation rate of conventional energy supplies (%/y)
M = maintenance cost (£)
T = lifetime of plant for auditing purposes (y)
E = annual energy produced by SWECS (kWh/y)
F = cost of energy produced by SWECS (£/kWh)

Note that E will depend on the wind conditions, the size and efficiency of the wind turbine, and on the efficiency of the load management control system. In these examples, we shall assume the following:

C_i = £10,500 for an 11 kW rated machine
       £35,000 for a 60 kW rated machine

B has three values: B_1 = 0%, B_2 = 10%/year
                    B_3 = 20% per year

I_1 = 5% per year

I_2 = 5% per year

M = 3% per annum of C

T = 10 years

E is the energy produced by the turbines rated for a wind speed of 9 m/s. The cut in wind speed is 4 m/s and the average efficiency is about 40% capture of the energy in the wind. In the tables the mean wind speeds are listed and a normal type of wind speed distribution is assumed.

147

A4.11 Simple method of economic assessment (zero price rise)

$$\text{Total money spent} = C_i (1 + MT + BT)$$
$$\text{Total energy produced} = TE$$
$$\text{Cost of energy } F = C_i(1 + MT + BT)/(TE)$$

(Equation 6.1)

Table A4.2 gives the value of F for the three different values of B at a flat rate and a range of values of E. Values of the mean wind speed are also given that are likely to produce E for the two machines. This simple method of assessment effectively assumes that increases in the costs of borrowing money and maintenance are compensated by increases in the price of conventional energy replaced, i.e. inflation is ignored.

Table A4.2 Cost of electricity produced, p/kWh, assuming zero price rise

	11 kW capacity SWECS				60 kW capacity SWECS			
Mean wind speed (m/s)	E (kWh/y)	B (0%)	B (10%)	B (20%)	E (kWh/y)	B (0%)	B (10%)	B (20%)
		Cost of electricity				Cost of electricity		
4	13,500	9.6	17.0	24.4	44,000	10.3	18.3	26.3
5	25,500	5.1	9.0	12.9	80,000	5.6	10.1	14.4
6	37,000	3.5	6.1	8.8	124,000	3.7	6.5	9.3
7	48,000	2.7	4.7	6.8	174,000	2.6	4.6	6.6
8	57,000	2.3	4.0	5.7	214,000	2.1	3.8	5.4
9	66,000	2.0	3.5	5.0	248,000	1.8	3.2	4.7

A4.12 Capital budgeting calculation

Instead of buying a SWECS for the initial cost C_1 the owner could have invested this money, received interest and had no maintenance charges. However the owner would have had to pay for conventional energy which would be expected to increase in price due to inflation and other causes. By buying a SWECS the owner has no fuel bills for the output of the machine, and the interest payments become less and less burdensome due to inflation. A method of assessing costs from these points of view is called present value discounted accounting (see Section A4.8).

The cost of energy produced by the two machines of our examples according to this technique is given in Table A4.3. Notice that cost is effectively reduced by this technique as compared with the simple method used for Table A4.2.

Table A4.3   Cost of electricity produced, p/kWh, using present value discounted accounting

	11 kW capacity SWECS				60 kW capacity SWECS			
Mean wind speed (m/s)	E (kWh/y)	B (0%)	B (10%)	B (20%)	E (kWh/y)	B (0%)	B (10%)	B (20%)
		Cost of electricity produced F (p/kWh)				Cost of electricity produced F (p/kWh)		
4	13,500	8.4	11.7	16.6	44,000	9.1	12.6	17.8
5	25,500	4.5	6.2	8.8	80,000	5.0	6.9	9.8
6	37,500	3.0	4.2	6.0	124,000	3.2	4.5	6.3
7	48,500	2.4	3.3	4.6	174,000	2.3	3.2	4.5
8	57,500	2.0	2.8	3.9	214,000	1.9	2.6	3.7
9	66,000	1.7	2.4	3.4	248,000	1.6	2.2	3.2

Note: These prices may be taken as the break even prices against conventional fuels assuming both the general inflation rate I and the conventional fuel inflation rate are both 5%.

Standard capital budgeting techniques give the total cash outflow C_O during the life of the plant discounted to present value:

$$C_O = C_i + 0.03 \ C_i \ \cdot \ \sum_{i=1 \ year}^{i=10 \ year} \left[\frac{1 + I_1}{1 + B} \right]^i$$

$$= 1.358 \ C_i; \ \text{for } B = 0\%$$
$$= 1.234 \ C_i; \ \text{for } B = 10\%$$
$$= 1.155 \ C_i; \ \text{for } B = 20\%$$

Applying the same technique to determine the discounted cash savings C_S on conventional energy:

$$C_S = E \ x \ G \ \cdot \ \sum_{i=1 \ year}^{i=10 \ year} \left[\frac{1 + I_1}{1 + B} \right]^i$$

$$= 0.119 \ x \ EG \ \text{for } B = 0\%$$
$$= 0.078 \ x \ EG \ \text{for } B = 10\%$$
$$= 0.052 \ x \ EG \ \text{for } B = 20\%$$

The break-even conditions occurs when $C_O = C_S$. Columns labelled B in Table A4.2 give the costs that the replaced conventional fuel should have for these conditions. If the present cost of replaced fuel is more than the figure in these columns, the wind turbine installation, on the basis of these assumptions, is likely to be economic.

The numbers given above are the basic economics of wind turbines. In many cases the initial purchase costs C_i may be reduced by offsetting against tax, or by benefitting from a number of grants available. It is very important to contact your Regional Department of Trade and Industry or Ministry of Agriculture, Food and Fisheries if you believe you might be eligible for an industrial or agricultural grant.

If the wind turbine is to be connected to the local area electricity supply system it is essential to contact the appropriate Electricity Board, who will advise on both the technical requirements and costs for the connection, and the effect of the wind turbine will have on your subsequent bills (see Chapter 5 also). Each Board has a variety of ways in which they charge consumers, and most operate different systems of charging. These aspects are discussed more fully in Chapter 5.

INDEX